PENGUIN BOOKS

A DANGEROUS PLACE

Marc Reisner worked for many years at the Natural Resources Defense Council. In 1979, he received an Alicia J. Patterson Journalism Fellowship and began research for *Cadillac Desert*, which was a National Book Critics Circle Award nominee, the subject of a PBS television series, and was included in the Modern Library's list of the Twentieth Century's 100 Best Nonfiction Books in English. He was also the author of *Game Wars*. Reisner died in 2000.

A Dangerous Place

CALIFORNIA'S UNSETTLING FATE

MARC REISNER

Penguin Books

PENGUIN BOOKS
Published by the Penguin Group
Penguin Group (USA) Inc., 375 Hudson Street, New York, New York 10014, U.S.A.
Penguin Books Ltd, 80 Strand, London WC2R 0RL, England
Penguin Books Australia Ltd, 250 Camberwell Road, Camberwell, Victoria 3124, Australia
Penguin Books Canada Ltd, 10 Alcorn Avenue, Toronto, Ontario, Canada M4V 3B2
Penguin Books India (P) Ltd, 11 Community Centre,
Panchsheel Park, New Delhi – 110 017, India
Penguin Group (NZ), cnr Airborne and Rosedale Roads,
Albany, Auckland 1310, New Zealand
Penguin Books (South Africa) (Pty) Ltd, 24 Sturdee Avenue,
Rosebank, Johannesburg 2196, South Africa

Penguin Books Ltd, Registered Offices:
80 Strand, London WC2R 0RL, England

First published in the United States of America by Pantheon Books,
a Division of Random House, Inc., 2003
Published in Penguin Books 2004

3 5 7 9 10 8 6 4 2

THE LIBRARY OF CONGRESS HAS CATALOGED THE HARDCOVER EDITION AS FOLLOWS:
Reisner, Marc.
A dangerous place : California's unsettling fate / Marc Reisner.
p. cm.
ISBN 0-679-42011-8
ISBN 0 14 20.0383 2
1. California—History. 2. California—social conditions. 3. California—
Environment conditions. 4. Disasters—California—forecasting. I. Title.
F861 .R43 2003
979.4—dc21 2002073375

Printed in the United States of America

For Margot and Ruthie

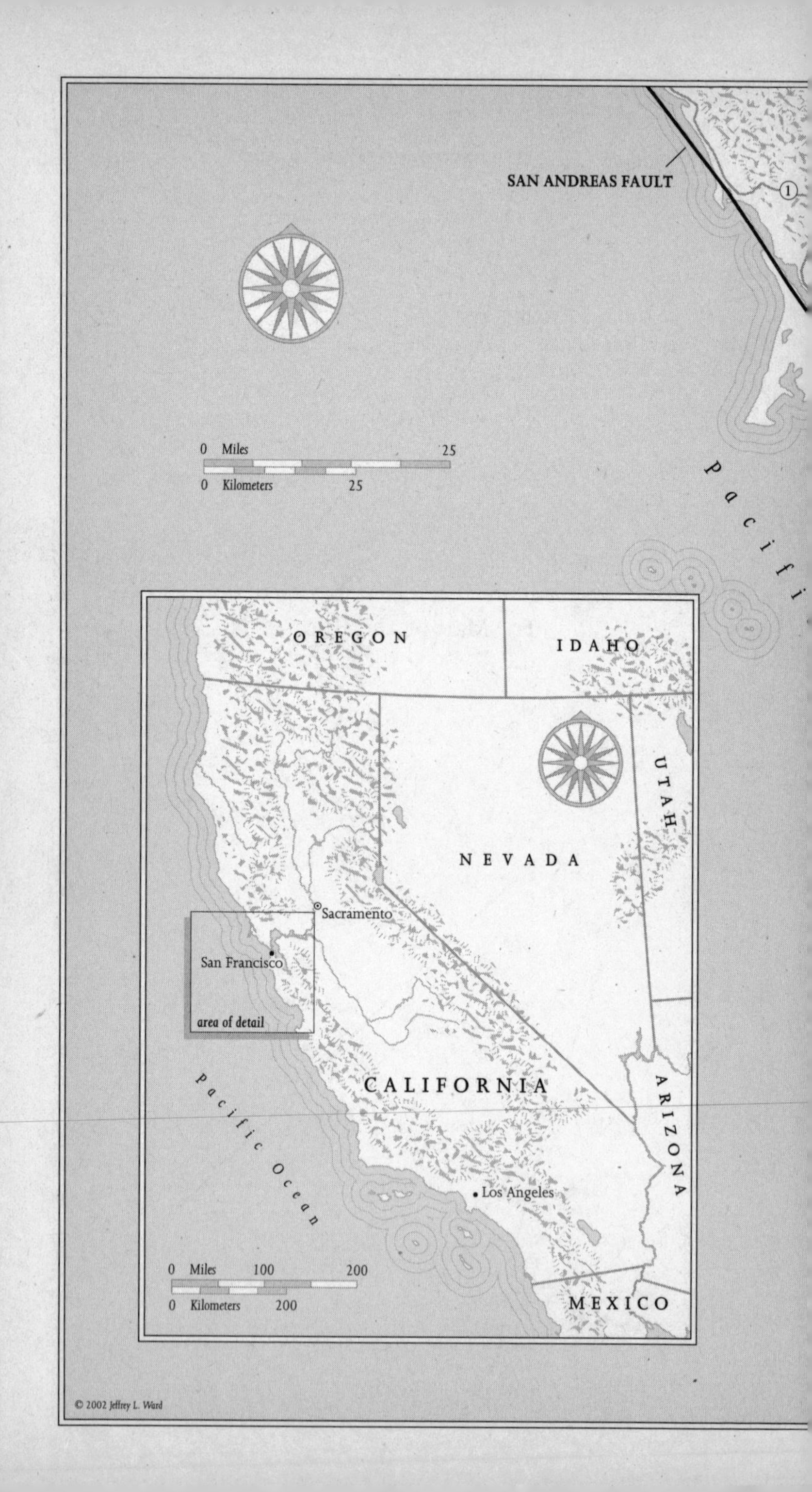
SAN ANDREAS FAULT
1
0 Miles 25
0 Kilometers 25
Pacifi
OREGON
IDAHO
UTAH
NEVADA
Sacramento
San Francisco
area of detail
CALIFORNIA
ARIZONA
Pacific Ocean
Los Angeles
0 Miles 100 200
0 Kilometers 200
MEXICO

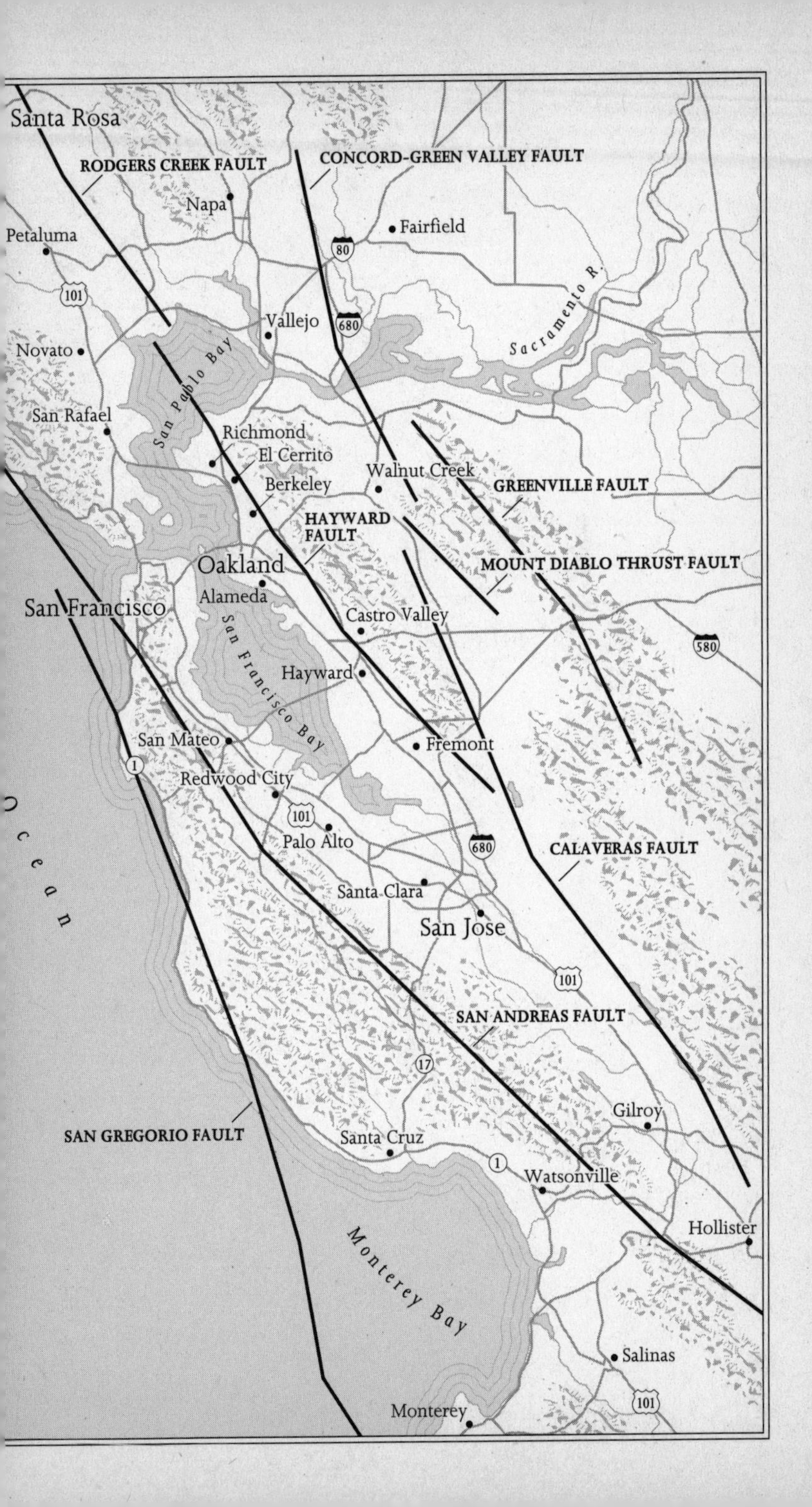

Santa Rosa
RODGERS CREEK FAULT
CONCORD-GREEN VALLEY FAULT
Napa
Fairfield
Petaluma
80
101
Sacramento R.
Vallejo
680
Novato
San Pablo Bay
San Rafael
Richmond
El Cerrito
Walnut Creek
Berkeley
GREENVILLE FAULT
HAYWARD FAULT
Oakland
MOUNT DIABLO THRUST FAULT
San Francisco
Alameda
Castro Valley
580
San Francisco Bay
Hayward
San Mateo
Fremont
1
Redwood City
Ocean
101
Palo Alto
680
CALAVERAS FAULT
Santa Clara
San Jose
101
SAN ANDREAS FAULT
17
Gilroy
SAN GREGORIO FAULT
Santa Cruz
1
Watsonville
Hollister
Monterey Bay
Salinas
101
Monterey

I

The most striking thing about modern California is not that it has transformed itself, in two long human lifespans, from a seamless wilderness into the most populous and urban of the fifty American states. Nor is it that one in two people living west of the hundredth meridian—which is to say, in half the landmass of the United States—now resides in California. Nor is it even that so many people are pouring in from everywhere that California is about to become the first state without a white—at least Caucasian—majority. All that is peripheral to the most fateful upshot of this state's century and a half of frantic growth: most of its inhabitants have settled, and will continue to settle, where they shouldn't have.

A swarm of Californians outnumbering the population of Texas has crammed into two small spaces: the Los Angeles Basin and the San Francisco Bay Area. These two regions, each roughly the size of the tiniest eastern states, now rank second and fourth among the nation's metropolitan areas. In an American West distinguished by its emptiness, they are gargantuan anomalies. Greater Los Angeles alone has more inhabitants than the ten other western states *combined*—states that have booming metropolises (Salt Lake City, Las Vegas, Seattle, Phoenix, Portland) of their own. It is the only conurbation in the West that makes the Bay Area, now nearly as populous as New York City, seem bucolic.

The appearance of such urban phantasms is mainly a twentieth-century phenomenon. In 1850, the world's largest city was London, with a population of two and a half million. Today, cities of five million people don't rate a second thought; there are some that hardly anyone has heard of, such as Tianjin, China. Greater Tokyo and Mexico City are both about as populous as Canada, but no one seems willing to say that their growth must stop—that they have grown impossibly, unmanageably huge. That is because, somehow, such chaotic, crawling hives have managed to cheat breakdown and catastrophe; at least they have so far.

California's two great megalopolises have done better. Despite the disappearance of almost two hundred thousand defense-related jobs, despite racial polarization and a creepily pervasive criminal violence, despite the departure, in the mid-1990s, of *both* its National Football League teams, Los Angeles remains one of the three wealthiest urban economies in the world—the regional economic product surpasses Africa's—and a magnet for tens of thousands of new emigrants each year. The Bay Area grows just as inexorably, if less famously. In the last twenty years of the twentieth century, its population soared from three million to six and a half million people; we're now twice as populous as Oregon, having absorbed, in the past twenty years, a population equal to that of Oregon.

Both Los Angeles and the Bay Area are the durable demographic wonders of American history: they exploded into existence, each at different times, and raced away from there. But does it make sense for so many people to inhabit them? From one perspective, nature's, the answer is a deafening no. In the first place, neither region has any water. There is some, but the local resources—some aquifers and seasonal streams—are enough to sustain hundreds of thousands in the long run, not many millions. Unlike Cairo—another huge desert anthill, but one that sits beside the Nile—Los Angeles and the Bay Area have had to import water over seemingly impossible distances and across rugged, desolate terrain. Los Angeles' aqueducts, viewed through telescopes from space, have given astronauts pause. If the contrived flow of water should somehow just stop, California's economy, which was worth about a trillion dollars as the new millennium dawned, would implode like a neutron star.

Building impressive waterworks to sustain humans in drought-prone places is not exactly new. Babylon and Mesopotamia and Ur did it, and so, even more impressively, did Rome. In fact, the majority of great civilizations in antiquity were in arid or semi-arid regions, and water, in one way or another, is linked to the demise of most. But an abject reliance on imported water isn't the Bay Area's, or the South Coast's, main vulnerability; it's the fact that they are astride one of the most violently active seismic zones in the world. The real estate and public infrastructure that's on top of California's major fault zones is valued in trillions of dollars—not billions. The economic loss from the next great earthquake (the last occurred in San Francisco in 1906 and in the Los Angeles area in 1857) might surpass the ten most expensive disasters in the United States history—might surpass all of them *combined*. Death and suffering, breakdown and paralysis, will be like nothing we've ever seen.

The damage from such a quake could, of course, be that much more severe if it occurs near the end of the seasonal drought that California endures each year—when it hasn't rained in months. (In 1923, the Great Kanto earthquake, whose epicenter was in Tokyo Bay, killed a hundred and forty thousand people and set off firestorms that incinerated much of Tokyo and Yokohama. In Japan there is year-round precipitation.) The odds are respectable that Los Angeles or the Bay Area will be struck during California's long seasonal drought, when the whole surrounding landscape is primed to burn. The water supply systems will be damaged, perhaps even drained; when that happens, after fires are ignited—as they will be—a lot of what has survived the ground shaking will burn.

So California snubs nature—what else is new? Since humans progressed from hunting and gathering, most human activity has become in some sense unnatural. A dam mocks the natural order; so does a cornfield; so does antibiotic intervention against disease. But among civilizations that have overextended themselves, California, rather than settle its human hordes where its water is and earthquake zones aren't, has done the opposite.

Carey McWilliams, perhaps the sharpest critic of his native state that California has produced, wrote a book called *California: The Great Exception*—an alien nation inside the American fold. California's exceptionalness began before it was an American state; its whole demographic history went against the continental grain. In most eastern and midwestern states, rates of growth went rapidly forward, peaked during the industrial revolution, then tapered off. Even other western states, with climates and landscapes more inhospitable than California's, grew quite rapidly from the moment they were settled by Europeans. In California, the Spanish made landfall nearly a century before the English landed at Plymouth Rock, but two hundred years later the whole population (excluding Indians) could have fit inside a concert hall. That was the first iconoclastic chapter in the state's demographic history—an extraordinary period of almost no growth. Then, as the growth of other states began to slow dramatically, California's took off with no brakes.

It began in 1769, with the arrival of Junipero Serra, a Majorcan priest in the Franciscan order. Serra, who has been morphed into a kindly old monk, was, like most of the friars, a man of remarkable toughness and comprehensive skills. He was also a self-righteous autocrat with a brutal temper and a penchant for self-flagellation—in the event of an erotic dream, he had at hand a nail-studded board hanging by his bed. Serra was adventuresome, easily bored, and intrigued by the country lying north of Mexico, which was unsettled mainly because explorers had not found any gold. When he was fifty-five years old, with a tepid blessing from the colonial government, Serra took a tiny party a thousand miles north, traveling on foot with pack horses, and, at a mission site near San Diego Bay, introduced California's first non-native culture: a despotic theocracy. Serra and the train of disciples that followed him for fifty years managed to found twenty-three missions, from San Diego north to the Sonoma Valley. And yet, despite all kinds of blandishments—land grants, cash subsidies, Indian slaves—the settlements that grew up around the missions had attracted just four thousand colonists by the time Mexico won independence from Spain in 1823.

The new Mexican government viewed the missions as an expensive liability, and, in the 1830s, ordered them secularized. The iron-willed priests had held these communities together; when they were gone, many Indians who had been lured into servitude ran off. Without their slave labor, the irrigation works that sustained the missions went slowly to ruin, and settlers who stayed generally turned to ranching. Some of the missions became ghost settlements, like the later mining towns. Those that held on to a few hundred settlers were at or near the coast, especially where the colonial regime maintained defensive presidios.

In the late 1830s and early 1840s, the majority of California's few thousand non-native inhabitants resided in three such places. Two of them—San Francisco, which was then called Yerba Buena ("good herb"), and Monterey—held obvious advantages as ports. The third was the most curiously situated. Two of Serra's advance men, Gaspar de Portolá and Father Juan Crespi, first reconnoitered there in 1771, but kept going and founded the next mission some miles farther north. Five years later, however, Felipe de Neve, the governor of both Californias, Alta and Baja, decided to establish a pueblo there anyway. Perhaps to make up for the shortcomings that induced the priests to move on, de Neve bestowed on the threadbare locale a grandiloquent name: El Pueblo de Nuestra Señora la Reina de los Angeles del Río de Porciuncula.

The Town of Our Lady the Queen of the Angels by the River of a Little Portion of . . . what? The pueblo was situated twenty miles from the coast, a long way by horse or foot. No decent harbor existed there, as at San Diego, Monterey, and especially San Francisco. The settlement was lost in a flat semi-arid plain, punctuated by some hills, that stretched forty or fifty miles inland before it collided dramatically—drastically—with a half-circle of high ranges that the priests named after various saints: Gabriel, Bernard, Jacinto, Ana. The highest of the summits, San Gorgonio Peak, is three hundred feet taller than Oregon's Mount Hood. The stark magnificence of these raw young mountains was a scenic attraction, to the degree that the friars were interested in scenery—reading their journals, one thinks they didn't give a damn—but the terrain made Los Angeles hard to reach. By land, you had to cross forbidding passes or coastal canyons thick with chaparral. In winter, arriving by ship, you risked wild Pacific storms. The mountains ringing the basin blocked the eastward movement of these same storms and wrung out their moisture—so efficiently that an occasional great storm could, in a couple of days, dump as much rainfall on Mount Wilson as Los Angeles sees in a typical year. The Los Angeles, San Gabriel, and Santa Ana Rivers, nearly dry at Thanksgiving, were sometimes miles wide on New Year's Day. In February, the peak of the brief rainy season, the nether reaches of the basin—what has become Long Beach, Culver City, Torrance, Carson—were sprawling marshlands covered with ducks and prowled by grizzly bears. Then everything began to dry up, and by October, just before the rainy season began again, a stray ember could prompt a holocaust. One local tribe, the Yang-Na, called the basin Valley of the Smokes.

It was a peculiarly inhospitable locale. Compared to Mexico or Minnesota—compared to most anywhere—the climate was wonderfully mild. The ocean got tolerably warm. The soil was good. Other than that, all was wrong. For years, for decades, Los Angeles' main attraction was its remoteness. By the 1840s, it had matured into a filthy, drowsy, suppurating dunghole, a social meltdown of Mexicans, Indians, Americans, various Europeans, Hawaiians, and a respectable number of freed or escaped slaves. Among its vernacular street names were Nigger Alley and Cunt Lane. Local enterprise amounted to cattle ranching, thieving, gambling, whoring, wine and brandy making, and a small fishing fleet (the distance from the coast gave the fish enough time to spoil). In the outlying portions of the basin, meanwhile, a leisure class of Latino grandees with huge land grants bestowed on them by the colonial government ran cattle over their fiefdoms, and invented entertainments. One dependable crowd-gatherer was the bear-and-bull bait, which involved a captured grizzly bear—there were many around as late as the 1860s—and a bull tethered together by their hind legs until one was chewed or gored to death. (The bear usually won.) Another amusement, *correr el gallo,* sent horsemen at full gallop to pluck half-buried roosters out of the ground by their necks. It was hazardous for the rider, the horse, and the rooster, too.

During the gold rush, when San Francisco turned into a roaring madhouse, the population of Los Angeles declined from six thousand to sixteen hundred—in one year. Everyone went north. Los Angeles had been California's largest town; now there were six tent settlements in the Sierra foothills that surpassed it, including one named Volcano, which is now extinct. Those who remained in Los Angeles were typically drunks, outlaws, or drunken outlaws, who read about their exploits—if they could read—in one of the town's two weekly newspapers. This item is from the *Star:*

> Last Sunday night was a brisk night for killings. Four men were shot and killed, and several wounded in

> shooting affrays. . . . A Yaqui Indian named Augustine del Rio murdered a Mexican because he refused to drink more whiskey after he went to bed. . . . On Sunday, the better the day the better the deed, a dispute arose between Nicho Alepas and a Negro named Governor Scholes, which resulted in the death of the former.

There was a smattering of pious farmer families around, especially Quakers fleeing bigotry and Mormons sent down by Brigham Young to investigate the region's agricultural possibilities. But until the 1860s, most food except meat was imported; the one important local crop, grapes, tended to become brandy or wine. "Los Angeles . . . is a vile little dump," wrote a God-fearing pioneer boy to relatives back east, "debauched, degenerate, and vicious."

As Los Angeles slid into a hole, San Francisco hooked itself to a comet. Its population in 1838 was roughly 700. Its population in 1848 was roughly 800. Its population in 1850, when the first reliable census was conducted, was 34,776. It had taken two years for an invisible population speck to become the fourteenth-largest city in the United States. By 1860 it was the eighth largest. It was ten times more populous than Los Angeles, which in 1848 had been nine times more populous than San Francisco.

San Francisco was something beyond the most explosive boomtown in history. It was the only nineteenth-century American boomtown among hundreds that never skipped a beat. It didn't decline, collapse, disappear from the earth—it just went on. That is not necessarily something to brag about, because the by-products of San Francisco's early years were horror and excess. During the gold-panning era, which went full tilt for only five or six years, miners and cavalry massacred Indians by the thousands. When they were bored killing Indians, they lynched Mexicans, blacks, and Chinese. Market hunters feeding San Francisco and the gold country towns and camps needed about eight years to wipe out the Central Valley's herds of antelope and elk, which some had compared to bison on the Great Plains; they also slaughtered waterfowl by the millions. Whole mountainsides—whole basins, like Lake Tahoe's—were shorn of virgin timber to erect San Francisco and dozens of mining towns.

The ugliest excess, though, was in mining itself. Gold panners turned over every last reach of streambed where a salmon might want to spawn, but that was nothing—the real harm to watersheds began with hydraulic mining. To mine gold hydraulically, you diverted a portion of a stream, or all of it, and ran it down a flume or canal engineered along a canyonside. (Most of this deliriously difficult and dangerous work was performed by Chinese, who crowded into California for miserable jobs that paid a dollar a day. In China, they got twenty cents a day.) After leading the captured stream down a long and gentle descent, accumulating hydrostatic head, you dropped it through an outsized metal pipe anchored to a steep slope. The outlet was coupled to a hose. At the end of the hose was a nozzle mounted on a heavy tripod—no human could have held on to it. The water, accelerated by the sharp drop and then constricted, which increased the acceleration, did not merely erupt from the nozzle. It exploded from it at a speed of a hundred fifty feet per second. So much water came out so fast that a hydraulic jet is supposed to have been able to decapitate a horse. It could also demolish a canyonside, liberating the gold sequestered in ancient raised streambeds. For every couple of ounces of gold, a thousand tons of rock, sand, gravel, and silt were set free, too. Hydraulicking was the most lasting carnage visited on California's landscape. It was finally outlawed in the 1880s—the first environmental victory won in the U.S. courts—after billions of tons of washed-down detritus had raised riverbeds as much as sixty feet, spread winter torrents wide, shut down inland navigation, silted out salmon spawning redds, and left eviscerating scars throughout the foothills. The lighter sediment ultimately drifted into Suisun, San Pablo, and San Francisco Bays, where much of it remains, shifting around on the tides, closing down marinas, causing peculiar miseries. In the 1980s, sediment washed down a century before helped demote an admiral who snagged his aircraft carrier on an unexpected bar.

From San Francisco's point of view, however, the hydraulic mining era simply refueled a sputtering gold rush and prolonged it another fifteen years. The last years of hydraulicking coincided with the silver boom set loose by the prodigious Comstock Lode—which was even farther away, in Nevada, but whose spoils ended up in San Francisco, too, making bigger fortunes than had gold. At the height of the silver boom, the first transcontinental train arrived in Oakland, a hop across the bay. San Francisco became the Las Vegas of the nineteenth century, living off rivers of wealth that poured in from elsewhere and the indulgence and excess that sudden money liberates. The city soon owned every bragging right in the transmountain West. It had the most banks, the most trade, the biggest mansions, the best hotels and restaurants (which offered amazing feasts of game), the most newspapers (twelve dailies by the 1860s), the largest fishing fleet, an awe-inspiring collection of brothels, the first important manufacturing industries (mining equipment, sugar refining, steelmaking, arms and munitions, apparel, shipbuilding, carriage making, chocolate), and the tallest buildings west of Chicago. It was a wonder of geography, accident, and mass migration, haunted only by scarcities of water and food.

In the 1860s, when the city's population passed 100,000, there was no place in America where so many people looked across so much water and had so little for their use. Even Los Angeles, where precipitation is substantially less—fourteen inches annually versus San Francisco's twenty—had a real river running through the center of town that, in flood, disgorged enough water for years, had there been some way to collect it. San Francisco had nothing you could call a watershed; a few rivulets sometimes qualified as creeks, but in summer they were usually dry. There were also some springs whose value was such that they were fenced by their owners and, in at least one instance, booby-trapped. Early entrepreneurs who had spurned or failed in the gold rush flourished in the water business, ferrying schooners back and forth between the city and flowing streams and springs coming off the Marin headlands and Mount Tamalpais. The water was transferred from ship to hogsheads, which were strapped to the backs of miserable mules; the mules were goaded up and down slippery streets, delivering water by the bucket at a price that corresponds now to ten or fifteen dollars per bath. (It was a spectacle such as this that prompted Andrew Hallidie, an animal lover and inventor, to conceive of the cable car.) Fresh water was so expensive that common people bathed in the frigid bay, slapping themselves with towels to remove the salt. Some washed their dishes there, too. Ultimately, because water storage was so chronic and there was so much wealth about, San Francisco achieved a further distinction: it was the first American city to build itself a Roman aqueduct. Technically a flume, it was constructed of redwood and brought water from Pilarcitos Creek, twenty miles down the peninsula. It was one of the biggest private engineering projects of the era, financed with fortunes made from the Comstock Lode. In the 1870s, the system's investors, incorporated as the Spring Valley Water Works Company, augmented its reliable yield by building several small dams, capturing enough flood flow to create a water surplus that would carry the city through the next fifty years. The dams formed reservoirs in the narrow valley from which the company took its name, a rift in the hills above the village of San Mateo. It was not then known that this peaceful place was the creation of violent wrenching along the San Andreas fault, directly below.

Early San Francisco's helpless dependence on imported food (except for fish and game) seems surprising, since the city is close to much first-class farmland. The Santa Clara Valley, before it became Silicon Valley, swept away to the south; the Napa and Sonoma Valleys were a short distance north. To grow much of anything there, however, you had to irrigate. In contemporary California, irrigation is so ubiquitous—nine million acres of cropland are artificially watered—that it is hard to imagine the despair of the first farmers who gave it a try. In the Sacramento Valley, the northern half of the Central Valley, many who did gave up after one winter season, when the river you could walk across in summer spread miles from its banks, chewing away irrigation headgates and silting the canals. In the winter of 1862, the wettest on record, the whole Central Valley was a shallow body of water the size of Lakes Erie or Ontario, and almost every farm and farmhouse was drowned. All the rivers in this climactically schizoid terrain behaved the same way—they were usually too empty or too full. To irrigate you needed river regulation. River regulation requires dams. Few nineteenth-century farmers and mutual water companies had the capital or expertise even to think about building dams. Some brought water up from shallow aquifers, but that was difficult and expensive before the advent of cheap oil and electricity and turbine pumps. Others put hogsheads on mules and hauled them to lovingly tended row crops.

Because irrigation was such a daunting enterprise—and because California's first population wave had a distinctive disinterest in farming altogether—a drastic share of gold rush–era incomes went for food. Potatoes were imported from eastern states and Hawaii; they were doubly expensive because half didn't survive the trip. A lot of fresh fruit and vegetables came from Hawaii; the Mediterranean fruit fly hadn't arrived there yet to destroy most of the state's thriving agricultural industry. (Hawaii was also the destination of much of the gold rush era's dirty laundry; there were too few women in California to wash everyone's clothes, and immigrants who later became launderers were still building flumes and railroads, or panning for gold.) Rice came from the Carolinas and Louisiana. In mining camps, food prices, supplemented by bacon grease, induced heart failure. Fresh eggs went for a dollar apiece. Oysters, the miners' caviar, came on schooners from the Pacific Northwest and commanded twenty dollars a plate (several hundred dollars in modern money); oysters made a Willapa Bay oysterman named Espy the richest pioneer in Washington Territory.

Still, there was nothing the scurvy-haunted miners craved more, in their torrid summer diggings, than fresh fruit—a huge slice of watermelon, say. Peter Burnett, a Tennessee miner who became an early California governor, once watched a man unload a buckboard of melons before a gathered crowd that snatched them up at three to five dollars apiece. "Such times," Burnett wrote wonderingly in his journal, "were never seen before, and will never be seen again."

The man with the buckboard full of melons was named Zwart. Edwin Bryant, the first American mayor of San Francisco, had encountered him several years earlier, in 1846, along the lower Sacramento River, smoking forty-pound salmon inside his house—a driftwood hovel with a tule roof, which was pitch black. Zwart, who emigrated from Holland, had been a very young member of the Bidwell party in 1841, which made the first successful, and perhaps the luckiest, trek across the Sierra Nevada, over the treacherous 9,800-foot Sonora Pass. The passage of years and lack of practice had made him forget his native language, and he seemed unable to decide on a substitute. His utterances, Bryant observed, amounted to "a tongue (language it cannot be called) peculiar to himself and hardly intelligible . . . a mix of French, German, English, Spanish, and *ranchería* Indian, each syllable of a word sometimes derived from a different language."

The tules of which Zwart made his roof grew everywhere in what was then a swampy wilderness east of San Francisco

Bay and its outlying hills, where the Sacramento and San Joaquin Rivers meet. The Great Central Valley, which the two rivers traverse from the north and south, appears as flat as a billiards table—John McPhee described it as "planer than the plainest of plains"—but actually slopes imperceptibly toward the center, where the rivers converge after sucking up two dozen tributaries. In Zwart's time this region was barely above sea level, and outflow to the saline bays was constricted by the Carquinez Strait, a gap in the East Bay hills. Carrying most of the Sierra Nevada's runoff, the rivers slowed and pooled; runoff was further stalled by the narrowing of the strait and by tides pushing inland from the bays. The result, especially in winter and spring, was epic overflow. The rivers formed a half-million-acre marsh, braided with many sloughs. In Louisiana, sloughs are called bayous, and the Sacramento–San Joaquin Delta was the most convincing imitation of lower Louisiana that existed west of there. The low islands between sloughs and river channels were a dwarf forest of tules, which look like cattails but grow like bamboo. Wherever overflow was less—on uplands a few feet above mean high tide—huge oaks and sycamores thrived, entwined by wild grapes. In all the world, only two major river deltas are this far inland, separated from the ocean by upthrust landforms; the other is China's Pearl River Delta, whence many of California's Chinese emigrated. The virgin California Delta was so vast, wild, and confusing—its sloughs meandered everywhere and led nowhere—that John C. Frémont ("The Pathfinder") lost a whole regiment in there for several days, and some who ventured in just disappeared.

But the Delta, as the unintelligible Zwart was the first to realize, made spectacular farmland if you drained the overflow, cleared the tules, and, during the rainless summers, kept its whistle wet. It was the one place where irrigation was simple, cheap, and relatively foolproof. Water flowed everywhere all year long—you didn't have to store it behind dams or lead it around in long, engineered canals. Along the rivers and sloughs were low berms formed by the sediments left by floods. They were natural levees. When you cut through one the overflow ran out; when you patched the cut the island stayed dry, at least in summer—at least in the summers of most years. The drying tules were easily torched. Then you seeded some ground, ran irrigation water over a berm through a length of hose, and watched how your garden grew. Zwart began with a few acres of melons in 1849, tending them all summer long as armies of miners tromped, sailed, and paddled by on their way to the foothills. The typical placer miner made four dollars a day, when he worked; three thousand dollars in nuggets and dust amounted to a respectable year. Zwart's first crop of melons earned him thirty thousand dollars. His income was greater than Zachary Taylor's, the president of the United States.

Zwart's success inspired imitators in a rush. The most ambitious hired Chinese workers to raise the heights of natural levees as added protection against spring floods (the Delta could yield delicious, and lucrative, produce by May, if it wasn't wiped out by a flood). Irish farmers driven to California by the potato famine soon made potatoes a famous Delta crop; a German named William Henry Myers got a yield of a million and a half pounds, he claimed, from a forty-five-acre spread. Whatever you planted grew. The Delta became a botanical gold mine; in the 1860s and 1870s it fed most of the state. Emigrants who drifted home after a California sojourn—to Massachusetts, or to Australia—discovered that Delta asparagus was already famous there. Varieties of corn, onions, and melons flourished; so did beans and squashes and peaches and apricots and celery. *In mendelium delirium*, growers tried rice, hemp, jute, chicory, spearmint, citrus, cranberries, mulberries, peanuts, ramie, pomelos, even sugarcane. In the West of the 1860s, only the Willamette Valley in Oregon and the Platte River Valley in Colorado approached the Delta as a nexus of concentrated agricultural wealth. Absentee owners tilled and reaped from mansions atop Nob Hill. Others built grand architectural curiosities alongside their fields, merging Italianate, Shingle, Eastlake, and any other style that happened to be *du jour*. Many of the grand homes are gone now or survive as creaky old wrecks, but the grandest of all, the River Mansion of Louis Myers, the Pear Magnate—he was the son of William Myers, the Potato Magnate—is still there, a fifty-room extravaganza on Grand Island's Steamboat Slough. Myers had spent three hundred fifty thousand turn-of-the-century dollars on his Italianate folly before competition drove him into foreclosure. His architects had already built him a small theater, a formal ballroom, and a bowling alley, and they weren't done.

Crops grew sporadically in the Delta for two reasons: long summers of slightly ocean-cooled sun, and the high organic content of the soil, which is essentially peat, a gift of many thousands of years of tule growth and death. In some parts of the Delta, the virgin peat stratum overlying basement muds was fifty feet deep, a five-story frosting of pure humus. Unfortunately, the high-octane nature of Delta soil comes with a fatal drawback. Its carbon content is so great that, when exposed to sun and air, it oxidizes. The natural tule cover had acted more or less as a suffocating shade, protecting the fragile underlying rot against the elements. But when the cover was gone and the fields were constantly plowed, the top few inches disappeared every year—volatilized or blown away. Farm machinery, which grew heavier over the decades, compacted what was left; peat soil is not just oxidable but fluffy, full of air. As a consequence, the Delta, from the moment large-scale farming began, started to subside. Leveed islands that were slightly above sea level in the 1850s were several feet below sea level by the turn of the century. They were ten to twelve feet down by the end of World War II. Today, almost all of the Delta—a region half the size of Delaware—is, like much of Holland, a netherland. The deepest acre, on Brannan Island, has in the past century and a half sunk to twenty-one feet below sea level.

However, the farmland is so valuable, and its owners so comparatively wealthy, that the Delta is constantly given new leases on life. Its terminal inundation is averted with levees—newer, taller, and broader levees. From decade to decade, the levees rise as the land inside them sinks. Most are privately improved, but the Army Corps of Engineers, which maintains the levees along main navigation channels, has invested hundreds of millions of federal tax dollars in their protection and restoration. When one breaks—Delta levees have failed more than one hundred times—and an island fills, the corps barges in pumps and, over the course of a week or a month, removes the lake. Dredges arrive and the eroded gash is frantically filled and riprapped. Farming goes on, and so does the region's descent.

As the Delta continues to subside, however, the levees are

more prone to failure. They function now almost as dams, groaning under hydrostatic pressure created by megatonnages of water bearing down. When you stand in a Delta island—they are the preeminent suboceanic islands in the world—you feel the same sort of clamminess that you experience at the bottom of a dam holding back a few billion gallons of water. The propellers of passing boats are a few feet over your head. The levees, however, are not built like dams; most are crudely engineered embankments of sand and mud trying to keep big rivers and an incursive San Francisco Bay—a saltwater bay—from inundating all within. The rivers and sloughs that flow through the Delta are headed into the bay, unless some of their flow is captured by Brobdingnagian batteries of pumps that sit at the Delta's southern end. In that case the flow goes toward the San Joaquin Valley and Los Angeles. Both regions get half their water supply from there, and every hydrologist who understands the Delta has had the same waking nightmare: a sudden mass levee collapse that lets the ocean fill the Delta, forming a diluted saltwater sea in front of aqueducts that normally ship water down to twenty million people. The one rare event apt to cause such a calamity is a major earthquake on one of the Bay Area's several known faults.

By the time the 1880 United States census was taken, the population of California had reached 864,694. Three out of four people still lived within a hundred fifty miles of San Francisco, and the prevailing demographic pattern—a populous north and depopulate south—had no reason to change. The gold and silver booms were playing out, which forced more Californians to do what almost everyone in America did—farm the land. Some of the best farmland was south, in the San Joaquin Valley and the Los Angeles Basin, but each place, because of wildly erratic river flows, was exceptionally difficult to irrigate. The San Joaquin was also owned, in large part, by agrarian counterparts of the eastern robber barons—by people like Henry Miller and Charles Lux, who amassed a cattle fiefdom a million acres broad. (They tried to monopolize the water rights, too.) The Delta, on the other hand, was a flourishing cornucopia, the Napa and Sonoma Valleys north of San Francisco were full of vineyards, and the Sacramento Valley was Kansas with a mountain view: a plain of wheat.

Los Angeles still didn't amount to much as a city, but the basin was becoming one great orchard. Irrigation farming, by the early 1880s, was nearly the sum total of the economy. Winter-fruiting navel oranges had been imported from Brazil and summer-fruiting Valencias from Spain, and they thrived so well, in different seasons, that a great patch of the basin had been transformed into a fake subtropic idyll of 450,000 orange trees. There were also, according to the chamber of commerce's fastidious bean counters, 48,000 lemon trees, 64,000 apple trees, 33,000 walnut trees, various orchards of other kinds, and enough grapevines to support six wineries, one of which was supposedly the largest in the world. (Since Los Angeles' population in 1880 was 11,311 and its export trade was modest—the city finally had a railroad spur but still no port—it must have been full of alcoholics.) Besides the farming and ranching population, there was also a swarming merchant class staging a ceaseless flea market that Stephen Longstreet, a novelist and city historian, once portrayed like this:

> Business and merchandising added crates, boxes, kegs, and bundles to the earth sidewalks curbed by planks. Under awnings, loafers and whittlers "set in chairs," while clerks inventoried new shipments of dry goods and notions. . . . There was a bedlam of street peddlers offering drinks, fruit, buttons, or spot remover, and snake-oil and medicine showmen at their pitches, bootblacks working on dusty or muddy boots. . . . The town was a strange mixture of those who were up and doing, the promoters, the yea-sayers, and those who had been defeated by drought, bad crops, debt, bad luck, illness, and foolish investments, merely existing in broken boots showing obscene toes.

Around the corner from this weird little scene was the most outlandish, explosive, and self-destructive episode of growth ever experienced by an American city, and for all anyone knows, by any city in the world. It lasted about three years, and began, more or less, when the Atchison, Topeka and Santa Fe Railroad's first southern-routed train wheezed into Los Angeles in the fall of 1885, opening up competition with the Southern Pacific, which had the link to Oakland. Besides its track and rolling stock, each railroad empire was composed half of debt and half of land awarded by the government for building track: the faster it filled its land with people, the faster it retired its debt. Ten days after the Santa Fe reached Union Station, it cut the straight fare from Mississippi Valley embarcation points by five dollars. It was the opening ante in a game of lowball poker that resulted in a suicidal fare war. The Southern Pacific retaliated with an eighty-five-dollar one-way ticket—ten dollars less. The Santa Fe went down to eighty dollars. The Southern Pacific bid seventy. Sixty dollars, fifty dollars—"Twenty-five dollars to Los Angeles!" yelled the Santa Fe's advertisements. For several weeks the fares on both railroads held at fifteen dollars. Then, in the spring of 1886, midwesterners woke up to a late blizzard of doorstep flyers: KANSAS CITY TO LOS ANGELES FOR A DOLLAR!! The dollar fare lasted one day, but even as the war began its surcease, almost every train bound for Los Angeles, from British Columbia to Washington, D.C., was filled. It was not a population wave; it was a population tsunami. The migrants, although emanating mainly from the Middle West, were not in the main farmers. If they were, they didn't come to farm. They were adventurers, misfits, merchants, shysters, failures, and hustlers—a great, mad, speculative rush of greedheads and opportunists.

It was a Ponzi spree. One-acre lots near the original pueblo that had cost fifty dollars were subdivided into five new lots and sold, days later, for three hundred dollars apiece. Local dentists, waitresses, hotel bellboys, policemen, and mechanics made swan dives into real estate. Land sales offices in the corners of fruit stands were rented out for a hundred dollars a month. Speculators wandered around cheerfully, ornamented with wads of bills. Owners of undeveloped land were practically chased down streets.

In 1884, the estimated population of Los Angeles had been 12,000. Thirty months later it was 100,000. A city directory was published; it was revised two weeks later with five thousand new names. Groves of oranges and lemons were removed to make way for flimsy homes of uncured wood. Vineyards were torn out and replanted in adjacent streets, which were still dirt. The frenzy blotched across the great, blank expanse of the basin like poison oak; the farther it went, the more inflamed it got.

Here is one full-page newspaper advertisement:

BOOM! *BOOM!*

ARCADIA

BOOM! *BOOM!*

Here is another:

HALT! HALT! HALT!

Speculators and Homeseekers, Attention!

$80,000—Eighty Thousand Dollars—$80,000
Sold in a Day at Marvelous
McGarry Tract

Here is a third, one word on one page:

RAMIREZ!

Here is a personal favorite:

Go wing thy flight from star to star
From world to luminous world, as far
As the Universe spreads its flaming wall—
One winter at Vernon is worth them all!

Where there was no there, there, suddenly there was. A man named Monroe founded Monrovia. He did it by purchasing a few acres and building himself a house, which he called a town. The land was barren. Above it was the mile-and-a-half-high, nearly vertical wall of the San Gabriel Range, some of the youngest mountains on earth, which sent down floods, mud, gravel, trees, thousand-ton boulders, and uphill homes during torrential winter storms. Monrovia lots measuring fifty by one hundred fifty feet sold out at a hundred dollars a mistake. Another man bought a piece of outlying nowhere, named it after Glendora, his wife, and in one day sold off three hundred lots.

Whittier is the coming place! It will dwarf Monrovia and eclipse Pasadena! Nothing can stop it! The Quakers are coming in from all over the place!

THIS IS PURE GOLD ! ! ! !
SANTA ANA,
The Metropolis of Southern California's Fairest Valley!
Chief Among Ten Thousand, or the One Altogether Lovely!
Beautiful ! Busy ! Bustling ! Booming !
It Can't Be Beat !
The Town Now Has the Biggest Kind of a BIG, BIG BOOM !
A GREAT BIG BOOM !
And You Can Accumulate Ducats By Investing !

Alosta, Gladstone, Beaumont, Arcadia, Glendale, Raymond, Lamar, Burbank, Rosecrans, Bethune, Mondonville, Olivewood, Oleander, Lordsburg, Happy Valley, Busy Vista, Ivanhoe, Alta Vista, Nadeau, McCoy ("The Real McCoy!"), Bonita, San Dimas, Ballona, Ontario, Walleria, Ocean Spray, Broad Acres, Glendora. This is a fair portion of the modern nomenclature of suburban Los Angeles, all coined in a few months. So many towns were founded that the founders ran out of wives and children after whom they could name their towns. Azusa, a place that still exists, the smoggiest city in the country, was created from the first and last letters of the alphabet, followed by U.S.A. "Peerless Long Beach . . . is not a new settlement, but a prosperous town of 2,000 people. The hotel will be doubled in size, with a billiard room for ladies, streets will be sprinkled, and a bathhouse, with hot and cold water, will be built."

The advertisements and the barkers promised subterranean water, iron ore, silver, gold, perhaps even oil (there actually was oil). Elephants and giraffes, lions and tigers and monkeys—animal refugees from a stranded circus—were drafted into salesmanship by one developer. There were smorgasbords, lotteries, raffles, and fifteen-piece bands. Naturally, there was fraud. Fifty thousand dollars worth of "view" lots were staked near the summit of Mount Wilson; they were all sold, not having been viewed. Lots advertised as having "water privileges" were also gobbled up—some were in streambeds. But anyone who entered the pyramid at the bottom rose even without resorting to tricks. There was land eighteen miles from Los Angeles city center selling for one thousand dollars an acre in 1888 that had sold for one dollar an acre in 1882.

The boom and the hoopla that fueled it were not new to Los Angeles; the boom mentality was merely more lunatic, the hoopla more preposterous. During earlier decades of nonentity the forsaken city had watched its fortunes rise and thud in a cattle boom (ended by drought), a wool boom (ended by competition), and a silkworm boom (ended by the revocation of mulberry tree subsidies when it was learned that Chinese silkworms didn't like American mulberry leaves). There had been an earlier speculative boom, too, which ended tragicomically. For some reason, everyone seemed to believe that this boom—this Big, Big BoooooooooooooooooooooooooooooooooooM—would last. It wouldn't just last—it would get *bigger*. Philip D.

Armour, the sausage baron who had started in California but relocated to Iowa after the cattle boom expired in drought, had kept his hand in basin real estate and, like most of those with much to gain, was orotund on the subject of growth. "This is merely preliminary," he proclaimed in the late spring of 1888, "to a boom that will outclass the present activity as thunder to the crack of a hickory nut."

The nut cracked about a month later, like an ostrich egg. The banks that had sustained it, often with usurious five-day notes, caught a chill and stopped lending money for any transaction outside the heart of the city; the loans they still offered were based on the "ancient value" of land—the value in 1883 rather than 1888. The bubble had grown so immense that the burst was spectacular; it slobbered over everyone. The president of a bank and the publisher of a newspaper disappeared into Mexico. The preacher at what had been Los Angeles' best-attended church departed with the life savings of several of his parishioners. Some laid-out towns never materialized; others—Monrovia and Glendora—were repopulated by coyotes and jackrabbits. Orchards that had fallen into the hands of waitresses and policemen and clerks went to ruin. By the end of 1888, the assessed valuation of Los Angeles County dropped to $20 million from an earlier assessment of $63 million. (More impressive than that depreciation was the fact that, before it occurred, the county transacted—in one year—about $100 million worth of hyper-inflated real estate; it had sold off pieces of itself for five times more than the whole of itself was now worth.) By 1889, trains arriving in Los Angeles were mostly empty, and trains leaving were full. For about two years, Los Angeles' population declined at a rate of three thousand people a month. Late in 1890, the city, although still a city, was half as populous as it had been three years before.

Los Angeles and environs now had one final opportunity to evolve into the sort of place that its circumstances seemed willing to tolerate. It could grow, in an orderly manner, into a medium-sized city—like Dayton or Topeka—surrounded by some outlying towns, dryland *ranchos,* and a reasonable acreage planted in irrigated crops. There was enough sustainable water to permit such a peaceable realm; there was not enough for two thousand square miles covered with humans or irrigated fields. But three forces ruined the opportunity as quickly as they could. One was the Santa Fe Railroad, which had started the whole *opéra bouffe* by launching the fare war. Another was the *Los Angeles Times*. The third and most successful instrument was the Los Angeles Chamber of Commerce.

Harrison Gray Otis, the publisher of the turn-of-the-century *Los Angeles Times* and the avatar of southern California's growth, was an early Los Angeles archetype. He was worse than an archetype; he was a caricature of an archetype. Like a number of southern Californians destined to achieve sultanish power and wealth, he arrived more or less a failure, in middle age. He had started out as a journeyman printer in Ohio; then he enlisted on the Union side in the Civil War and earned his only bragging rights—a brace of medals and the rank of captain. Although Captain Otis admired, as he declaimed a hundred million times, "Hustlers! . . . men of brains, brawn, and guts!" he spent the years after the war chasing down sinecures. He was rewarded with a brief assignment as U.S. government agent on the Seal Islands, in the Bering Sea, where he chased after poachers who were decimating the pinniped herds. Otis bore a curious resemblance to a walrus—he was big and blubbery, with a perpetual scowl and an Otto von Bismarck mustache and goatee—and had a disposition that would have put a polar bear to flight. He was vituperative, choleric, hard-hearted, self-righteous. His opinions were unfailingly strong, and the tone with which they were expressed was unfailingly loud. A lover of war, and a fireside military strategist, he referred to his mansion as "The Bivouac." His weekend retreat was "The Outpost." Otis's long-suffering staff at the *Times* was "The Phalanx." His politics were antediluvian; Otis was a reactionary even by the standard of the Robber Baron era. After he had built the *Times* into the most bitterly anti-union newspaper in the West, and was anointed organized labor's Public Enemy Number One, he celebrated by having a miniature cannon mounted on the hood of his touring car. It fired live rounds.

Like some of his modern counterparts in the Republican Party, Otis stuffed his sermons with biblical references, but his real religion amounted to the open shop, capitalism (the more cold-blooded and rapacious, the better he liked it), and growth. After his half-frozen family finally convinced him to give up the Seal Islands post, he moved them to Santa Barbara, which was, and still is, a redoubt of the nation's leisure class. Otis despised the trust-fund rich, and even more he despised a city that resisted growth, as Santa Barbara always has. After a brief residency there, the Otises moved to Los Angeles, where the Captain invested most of what he was worth in a struggling weekly published by an eastern financier named H. H. Boyce. Publishing was as primitive as the town itself: Boyce's printing plant was still powered by water delivered through a *zanja*—an irrigation ditch—and its press runs were sometimes stopped by fish and debris clogging the intake. Otis's machinations and sclerotic nature soon drove Boyce away; he founded a rival newspaper that his former comrade referred to as *The Daily Morning Metropolitan Bellyache*. Otis's editorials were so vitriolic, his news stories so flagrantly slanted (he often said that he considered objectivity "a form of weakness"), that he drove his enemies, who were legion, to oratorical summits. If he had an archenemy, someone powerful who loathed him as powerfully as Otis loathed him, it was Governor (later senator) Hiram Johnson, who once offered this opinion of the Captain during a campaign speech:

> In the city of San Francisco we have drunk to the very dregs of infamy. We have had vile officials, we have had rotten newspapers. But we have had nothing so vile, nothing so low, nothing so debased, nothing so infamous in San Francisco as Harrison Gray Otis. He sits there in senile dementia with gangrene heart and rotting brain, grimacing at every reform, chattering impotently at all things that are decent, frothing, fuming, violently gibbering, going down to his grave in snarling infamy. . . . My friends, he is the one thing that all Californians look at when, in looking at southern California, they see anything that is disgraceful, depraved, corrupt, crooked, and putrescent—THAT is Harrison Gray Otis!

Like many poor representatives of the human race, Otis had serious talents. He was a good judge of character—of heartless character, anyway. Early on, he was prescient enough to hire as his circulation manager a young man named Harry Chandler, who controlled most of the city's independently owned circulation routes and could ensure that the *Times* was delivered on schedule and other newspapers were not. When he threatened Otis with this trick, the publisher's fury yielded to admiration, leading to one of the nastiest partnerships in the history of capitalism.

Chandler was another Los Angeles archetype. Like many in the first wave of emigrants, he arrived in poor health. Chandler suffered from a serious lung disorder (which was caused or perhaps just aggravated when he dove into a vat of starch on someone's dare), and he moved from New Hampshire, where he was enrolled at Dartmouth College, in hopes of getting well. He eventually did, after picking and selling a doctor's fruit crop. He earned three thousand dollars in a single summer—more than most of the forty-niners made in a year—and funneled his savings into the circulation routes. Chandler bore not the slightest resemblance to the ferocious-looking Otis. He had worked as a child model, and his choirboy countenance aged into that of a minister. But he was just as acquisitive and ruthless, and, like Otis, took great pride in his flawed character.

Chandler and Otis did not make their real money in publishing; mainly, they used their newspaper—whose editorials praising growth and development were outrageous even by the standards of the day—as a propaganda tool to reap a fortune from real estate. The Chandler family land empire ultimately comprised at least 300,000 acres in California and another 860,000 acres in northern Mexico, where Harry grew cotton by sucking the lower Colorado River dry and jeopardizing his own city's future water supply. Their empire had its beginnings in the land boom of the 1880s, which exploded soon after he and Otis acquired complete control of the *Times*. Otis's bellicose editorials flogged the speculation along, and after it all collapsed he sugar-coated the carnage. "The 'busting of the boom' became but a little eddy in a great stream," soothed the *Times* a postmortem; "the intermission of one heartbeat in the life of . . . the most charming land on the footstool of the Most High . . . the most beautiful city inhabited by the human family." But treacly epiphanies didn't stop hordes of people from moving away, and the *Times* was losing revenue; it had become mightily dependent on land-sales advertising. The only answer was to lure new conscripts faster than the burnt-out phalanx was leaving.

The Santa Fe Railroad, now running empty trains, had a similar idea. Otis wanted heavy industry, a port—anything that would make and keep the city populous and inflate the worth of his properties. The Santa Fe was more interested in farmers, who added social stability and a lot of tonnage to fill its trains. Whatever worked. Everyone knew that the local economy could no longer be created from flea markets, Ponzi schemes, brandy, and silkworms.

The result, the Los Angeles Chamber of Commerce, was a revelation. A chamber had existed since 1873, but it was as drowsy as the town had been then. When it was invigorated, which is much too mild a word, by Otis and company in 1888, it was decided that the first order of business was to drain away the solid citizens of the Middle West. Luckily, the whole midwestern region was then in the midst of serial catastrophe. The Great White Winter of 1886 and 1887 was the fiercest on record and wiped out much of the High Plains' cattle industry. It was followed by several years of drought, which culminated in the financial Panic of 1873. A great many shell-shocked midwesterners were ready to leave; they simply had to be told where to go. The chamber began filling the Santa Fe's trains with two types of propaganda: pamphlets and oranges. Its first piece of literature, *Facts and Figures about Los Angeles City and County,* was a thirty-six-page confection that mixed up fruit, sunshine, snowy peaks, sandy beaches, cheap land, religion, irrigation, Creation, and year-round warmth. Among the facts and figures wronged by *Facts and Figures* was the population of Los Angeles, which was 40 percent smaller than it said, and shrinking. It rhapsodized about "an abundant subterranean water supply, which was not previously known to exist"; because of the magical aquifers, "Southern California can no longer be referred to as a semi-arid region . . . sources are sufficient to irrigate 600,000 acres of land." (With just half that amount of land irrigated, the basin's aquifers and artesian wells were going dry only twenty years later.) That irrigation was even required for many crops was "a mistaken idea," despite the fact that, in a typical year, the region went rainless for at least six months—and those were the summer months, the growing season for most crops.

Facts and Figures and a sequel called *The Land of Sunshine* were written by Harry Ellington Brook, an editorial writer who Otis loaned from the *Times*. Their success inspired a future avalanche of propaganda under Frank Wiggins, the chamber's president for almost thirty years. Wiggins came across as a Quakerly Parson Brown but had the mettle of P. T. Barnum; a business college was founded in his name. Seventy-five thousand copies of *The Land of Sunshine* were printed and rushed, in mail cars also stuffed with citrus fruit, to Chicago on the Santa Fe's fastest trains. The chamber had rented an exhibition hall there, for the practical reason that Chicago was the bottleneck through which most of the nation's westward migration squeezed, and many migrants arrived there still unsure where they planned to go. The fruit decorated a towering ziggurat that was the exhibition's centerpiece. Pamphlets lay around in mountains, and orators worked the crowds.

The Chicago exhibition was the opening drive in what may have been the most successful propaganda campaign ever conceived. Las Vegas is the one American city now doing anything comparable, but its advertising asks only that people visit; the chamber wanted people to stay. In its first two years, the steroidally reincarnated Los Angeles chamber published more than a million copies of forty different pamphlets and booklets. There was a pamphlet on celery growing; there was another on grapes; there was one on alfalfa. Special pamphlets were prepared for the Atlanta Exposition (25,000 copies) and for another one in Omaha (50,000 copies). Fifty thousand copies of a booklet called *Southern California* were mailed to various editors and journalists at eight thousand publications across the United States; each was accompanied by a notice that more free copies would be sent to "any friend." By 1901, some chamber propaganda pieces had supposedly been seen by one of every five adults in the country. The campaign marched overseas. In 1901, the Chamber commissioned a dazzling hall at the Paris Exhibition; it had already done the same in Guatemala, and in Hamburg. (The *San Jose Times-Mercury* griped that "[t]he average . . . mind conceives of California as a small tract of country situated in and about Los Angeles.") One early spring it brought three hundred members of the National Editorial Writers' Association to Los Angeles and gave each of them a news story they could pretend to have written simply by affixing a byline. It was remarkable how many of the nation's editorial writers were seduced by southern California's beaches: the article appeared, virtually unchanged, in at least two hundred fifty newspapers in the eastern states. The editors were in such a state of winter fatigue that few bothered to correct a sentence that began, "The day was bright and beautiful, neither too warm nor too cold. . . ." (The temperature on the day referred to reached one hundred degrees; it was Los Angeles' hottest April reading in twenty-nine years.) But even such heat wouldn't have bothered the hordes of gold miners then converging on Alaska's Klondike region; the chamber distributed ten thousand copies of a land development brochure written expressly for them, in case they felt like bringing their fortunes down.

In 1887, the population of Los Angeles had reached 100,000. Two years later, it had fallen to 50,000. Ten years later, it had finally passed 100,000 again. This time, it kept going. The population grew and grew, and the rate of growth grew, too. By 1910, there were 319,000 people in Los Angeles—the population had tripled in a decade. It would double again by 1920, the beginning of the boom era prematurely forecast by Mr. Armour, the sausage baron—a sustained growth spurt when Los Angeles assimilated more people each and every year than during its first century of existence.

Although the chamber fired its propaganda rockets everywhere, it sent barrage after barrage into the Middle West. So many people migrated from Iowa that an Iowa reunion picnic staged at a fairgrounds outside the city in the late 1920s drew a crowd of 110,000, almost as many as then lived in Des Moines. (H. L. Mencken's first impression of Los Angeles was "double Dubuque.") Most of the transplanted midwesterners were farmers, and they continued to farm; they traded a few hundred acres of corn or wheat for a few acres of oranges or plums, and did just as well. Then they made minor fortunes selling their land to developers, and invested in more property. When the family circle back in Kansas learned how rich Uncle Horace and Aunt Ethel had become, they all wanted to emigrate, too.

Even though it was urbanizing rapidly, Los Angeles' geographic area was so majestic that, during the First World War, when the city had the population of Pittsburgh, it was still a city of orchards, vines, melons, and zucchini fields. As for the County of Los Angeles, it would reign unchallenged, until about 1940, as the richest agricultural county in the United States.

When the human influx was barely beginning, the chamber of commerce had already decided that transcontinental railroads weren't enough. The building boom demanded vast quantities of lumber, but the region was too arid for trees; some lovely fir and pine forests hugged the summits of the San Gabriel and San Bernardino Ranges, where precipitation was three times higher, but those forests had recreational value, and could never have built a city like the one materializing below. Lumber came cheapest by ship; so did most heavy goods. Los Angeles now needed a harbor—a real one, not the half-protected cove at San Pedro that was swept by dangerous rollers when winter storms barreled in from the south. (So many ships had sunk on their approach into port that, during prolonged rough weather, eastern goods bound for southern California were sometimes off-loaded at New Orleans and transferred to trains.) What the city was *really* after was federal subsidies, since it lacked—or so it said—the tax base to build such a monumental work on its own. It was the beginning of that conservative region's infatuation with federal largesse. In the spring of 1889, a private Southern Pacific express train arrived with six members of the Senate Commerce Committee on board, chaperoned by the usual complement of southern California boosters, experts, and orators, who had organized themselves as the Free Harbor League (the name was apt, since they wanted someone to build them a harbor for free). The group had been joined by Senator Leland Stanford himself, who was also president of the Southern Pacific. Senator William Pierce Frye of Maine, after viewing the harbor site and contemplating the cost of dredging and then building a miles-long breakwater, said it would be cheaper to move Los Angeles to where there was already a natural harbor—Maine, for instance. Frye meant it as a joke, but after it made the headlines the next day, the senator felt obliged to apologize. In 1891, his committee authorized an investigation by the Army Corps of Engineers, which soon gave its blessing to the San Pedro Harbor Project. (The corps had acquired a warm feeling toward public works projects, which offered more reliable engineering than did war, and would soon develop a mania about dams.) To ensure that it was its port and not San Pedro's, which was next door, Los Angeles simply redrew its borders through eminent domain, grew a long, narrow neck that reached to the coast—today, it is the Harbor Freeway corridor—and opened its jaws and ate the site.

With two transcontinental railroads and an expensively engineered harbor, rapid and continuous growth was inevitable. But no one foresaw pandemonium. Early in the century, the city's water department had projected a population of 400,000 by 1925; instead, it had 1,200,000 on its hands. Los Angeles' rate of growth was fifteen times faster than Denver's (and it was growing fast) and eleven times greater than New York's. In the twenties, the city gained 100,000 people a year. Then the Great Depression coincided with the Great Drought of the 1930s, and it absorbed 300,000 new migrants in eighteen months. The peaceable kingdom was vanishing; orchards were being razed; Los Angeles and its satellite towns rushed at one another in a suburbanizing blur. Smog alerts were just a generation away.

Some of the frenzy was due to serendipity—or to bad luck, depending on one's view of growth. The motion picture industry, still in its infancy, moved to Hollywood, a dusty little suburb, early in the century; it was too difficult to shoot in the dark and grimy reaches of Manhattan's Lower East Side, where a handful of Jewish filmmakers had inaugurated the business. They could have gone to San Francisco, or to Colorado, or even to Israel (where the climate is much like L.A.'s), but they went to southern California, and the industry grew up so fast that, by the twenties, it was the second-largest employer in town. In 1892, Edward Doheny, an oil prospector, discovered a pool of crude beneath the intersection of Second and Glendale. By the turn of the century more than twenty-three hundred derricks had sprouted within the city limits, and the great Signal Hill strike was yet to come. All kinds of industries—steelmaking, rubber, automobiles, construction—migrated into a region with its own abundant reserves of cheap oil. The absence of strong unions—Otis and Chandler could claim much credit for that—made their decision easier. The outgoing traffic in oil, manufactured goods, vegetables, and fruit, and the greater incoming traffic in lumber—at times, a couple of dozen lumber ships might be waiting offshore—transformed Los Angeles into the nation's second-busiest port in twenty-three years. Aviation pioneers—Glenn Martin, Jack Northrup, T. Claude Ryan, Donald Douglas—were, like the film industry, attracted to the sunny weather and set up skunk works in the twenties that grew into huge aerospace companies during the Second World War. Like almost everything else, aerospace would eventually feed off the federal government—the dynamo behind the city's post-Depression growth. World War II and the Cold War and its occasional conflicts bequeathed full employment on the region for fifty years. (Vietnam was a war waged with bombers and fighters and missiles manufactured in California, the state where the anti-war movement found its initial strength.) The culmination of the defense-spending binge was the military extravaganza staged in the 1980s by the Reagan administration. But even in the mid-1990s, after the Soviet Union had disintegrated at a cost of two hundred thousand defense-related southern California jobs, and after greater Los Angeles had fourteen million people and a reputation (at least among many Americans) as overcrowded, polluted, expensive, crime-ridden, and traffic-jammed, the region was still gaining people at a rate of twenty thousand a month.

Carey McWilliams, in *California: The Great Exception*, wrote that, "There was never a region so unlikely to become a vast metropolitan area as Southern California. It is . . . man-made, a gigantic improvisation." By contrast, few regions in North America were as *certain* to grow as populous as metropolitan San Francisco. Wrapped around the most expansive harbor south of Puget Sound, proximate to the Sierra Nevada's gold and to what became the world's most productive farmland, the terminus of the first transcontinental railroad, the Bay Area's growth was as foreordained as Los Angeles' was contrived.

In 1869, when San Francisco—a speck on the map twenty years earlier—was about to be revealed, by the 1870 census, as the eighth-largest city in the United States, the first train steamed into Oakland on the first transcontinental track. Now it was that city's turn. During rail's decades of glory, the 1870s through the Second World War, train traffic created so much Bay Area employment that in Oakland—which since the twenties has had the second-largest black population in the West—a third of the black wage earners worked for the Pullman Company. Gradually, the railroads were supplanted by highways, and the United States was not just the world's biggest car-builder but the world's biggest exporter of cars. Chevrolet, Willys, General Motors, and Kaiser erected assembly lines in or near Oakland, taking advantage of the port and the railroads they were sending into decline.

But nothing fueled Bay Area growth like war. Even though the First World War was mainly a land war, and the theater of naval combat was the Atlantic, a single Oakland shipyard employed 13,000 workers toward that war's end. Wartime activity, and money, played a central role in resurrecting an urban region that had been half-destroyed by the 1906 earthquake. Then came World War II, where the main theater of naval action was ultimately the Pacific. The impact of war production on the Bay Area was extravagant. In 1940, the Oakland industrialist Henry J. Kaiser, famous for having built Hoover Dam, knew next to nothing about shipbuilding, but with Roosevelt's Lend-Lease program enacted he approached the British government with an offer to build them cargo ships. Britain flew over some military emissaries late in the year 1940, who were taken by Henry J. to a bare mudflat along the Richmond shore and asked to imagine a big shipyard there—which they must have done, because Kaiser had a contract on his desk when they left. The shipyard was completed six months later, and began to crank out an initial order of thirty vessels. Meanwhile, San Francisco's Bechtel Corporation, a Kaiser partner in the Hoover Dam project, was approached by the U.S. Maritime Commission about a second new shipyard. Within twenty-four hours, on March 3, 1941, Bechtel had picked out a site, along the shallow Sausalito waterfront, a site that would require a tremendous amount of dredging. But the Marin shipyard had already laid its first keel by June 27, and the company was simultaneously building a town of apartments at the northern edge of Sausalito to house its thousands of workers. (Called Marin City, it was to metamorphose from Russian and Irish to black; the murdered rap star Tupac Shakur grew up there.)

By 1944, there were thirty shipyards operating around the bay, employing close to a quarter of a million workers. One hundred thousand worked at Kaiser's Richmond yard alone. According to Wayne Bonnett, a shipbuilding historian, the Bay Area's output of military ships nearly equaled the production of the whole Atlantic coast from Maine to the Carolinas. The demonic Kaiser led the way, establishing production schedules that seemed inhuman. Mainly, the Kaiser Richmond yard constructed Liberty ships, stolid cargo vessels that were 441 feet long and handled about nine thousand tons of matériel. One of them, the *Robert E. Peary*, was completed in four days and fifteen hours—a record that seems unsurpassable, like Wilt Chamberlain's 100-point basketball game.

There weren't nearly enough white males (the preferred labor force) in the Bay Area to build a couple of thousand big ships in fewer than five years, so the companies recruited women and even went south to recruit blacks, who faced discrimination in the Bay Area that was nearly as bad as anywhere else. Black workers also tended to perform the most difficult and dangerous jobs; at the Port Chicago yard, near Antioch on Suisun Bay, hundreds of black navy workers perished when a huge munitions load exploded.

During the Second World War, the federal government spent over $5 billion—about $60 billion in 1999 dollars—on war-related contracting in the Bay Area. The federal government, mainly the military, employed more people in the region than other levels of government combined—county, city, and state—and those ranks don't include the hundreds of thousands who gave this mass labor force housing, streets, sewers, electricity, water, entertainment, and food. Military activity swelled the Bay Areas census by 600,000 during the war years, even though much of its male population was going overseas. In fact, millions of young males from elsewhere in America—those hurled into the Pacific theater—caught their first rapturous glimpse of the Bay Area as they were shipped out, and many vowed that, if they survived, this was where they would move. They held true to their promise, married, built tract homes with government loans, and began to reproduce.

Meanwhile, the region's great universities were spilling out scientists and engineers—another offshoot of the war effort—who founded and staffed technology companies that would eventually turn tens of thousands of acres of

Santa Clara Valley orchards into a sanitary Ruhrgebiet known as Silicon Valley. Between 1940 and 1995, San Jose, the world headquarters of cyberspace, was transformed from an outsized village of forty thousand into a mesmerizing sprawl of a million people, with another million in its endless suburbs. It seems inevitable that Silicon Valley will sprawl all the way down to bucolic, seaside Monterey, seventy miles south, which shudders at the thought. It seems just as inevitable that, at the other end, Sacramento and the Bay Area will eventually link, forming a sanitary version of the New York to Washington corridor with a population vaster than Los Angeles' at millennium's end.

All of this happened, and is happening, effortlessly. The Bay Area never conspired to grow; it never advertised itself around the world; it had no real need for a chamber of commerce, *Facts and Figures*, exhibits at international expositions (San Francisco staged its own, in 1915 and 1939), or silkworm booms. It just grew. Unlike Los Angeles, it needn't have worried a jot about attracting people or industry. Its real problem was where to put them.

The environs of San Francisco Bay comprise no sprawling plain. Flatland—easily developed land—has always been scarce. Building on steep slopes was much more difficult in the nineteenth century than today, and the original settlement of Yerba Buena was worked into a sliver of shoreline squeezed between the bay and Nob Hill. When San Francisco's growth took off, development was forced uphill, but the cost of building on slopes and the inconvenience, in the horse-and-buggy era, of living on them were such that someone, at some point, decided that a better alternative was to create new land around the margin of the bay.

It wasn't difficult. You just excavated a hillside and dumped the spoil in the shallows, supplementing rock and gravel with dredged-up bay mud and anything else semi-solid—garbage, wrecked houses, a scuttled ship whose captain and crew ran off to mine gold. By the 1860s, a large chunk of what had become downtown—much of the area bayward of Sansome Street, where the shoreline had once been—rested on fill. Bit by bit, sometimes in increments of hundreds of acres, the bay perimeter became land. The same happened in Oakland, especially around the island of Alameda. It happened in Berkeley and Richmond. In 1915, San Francisco staged the Panama-Pacific International Exposition, its triumphant return from near devastation, along the stretch of bay out by the Presidio and Golden Gate. Some of the grand buildings were on a piece of ersatz land that had been a large tidal lagoon. Soon afterward, the whole fantasyland of exhibition halls (except for one neo-classical survivor, which is now the Exploratorium) was ruthlessly demolished, and the site metamorphosed into the Marina District, which has become a mecca for Financial District yuppies who work, sleep, network, and jog on what used to be water. The original airport, nearby, was moved to a larger and better site, a filled-in salt marsh on the other side of the peninsula, south of Candlestick Point. Each new runway at deliriously busy, forever-under-construction SFO (San Francisco International Airport) demands more fill; there is nowhere to expand except into the bay. Much of the Oakland Airport sits on ex-marshland or bay. The Bayshore Freeway, U.S. 101, which connects San Francisco and San Jose and replaced the old real-land highway, El Camino Real, would once have been a mile out from shore. Whole towns rose out of the former bay—Foster

City, an upscale Levittown, and Milpitas, and Alviso. Alameda Island doubled in size. Treasure Island, created as the site for a second exposition, the 1939 Golden Gate International Exposition, did not exist two years earlier.

By the 1960s, when a third of San Francisco Bay had already vanished and San Pablo Bay, to the north, was also disappearing fast, developers and their allies in the conquest of nature—the secular religion of the time, even here where the Sierra Club was born—seemed no longer content to just shrink these great joined estuaries. They wanted to get rid of them. A theatrical promoter and self-described visionary named John Reber conceived a colossus of a project that would have diked off a great portion of the bay, turning it into a freshwater lake. A large section that wasn't a lake would have become land. Someone else wanted to build a concrete platform over the South Bay and cover it with homes and . . . parks. The Southern Pacific, in its imperishable quest to monetize the 11,588,000 acres it had acquired through nineteenth-century federal land grants, wanted to level San Bruno Mountain, a thousand-foot rise north of the airport, which it owned, and dump the excavated mountain into the bay; with luck, there would be enough fill to reach Oakland. (Somehow, the Rockefeller family was involved in that scheme, too.)

As urban planning (it was actually called that) of this ilk was trotted out for public approval, the mood of Bay Area residents finally improved from docility to near violence. The Bay Conservation and Development Commission was created in 1965, and given enough authority to put an end to such schemes, if not such talk. But the potential consequences of creating so much land out of marsh and water were unrecognized for some time. Large commercial buildings and apartments, factories, airports, military bases, freeways, bridge approaches, ports—whole neighborhoods—now lie over semi-saturated pseudo-terrain in the heart of one of the planet's most active seismic zones. This is land in the loosest sense of the word. When stressed badly enough, it doesn't behave like land. Semi-saturated, unconsolidated soils slow down seismic waves broadcast by strong earthquakes. As the seismic energy is delayed, shaking amplifies (think of a hurricane that stalls). If the shaking is strong enough, trillions of soil particles lose their cohesiveness. Water within loose sandy soils increases in pressure, to the point where soil particles begin to float and the land liquefies. To appreciate the dynamics of seismo-liquefaction—which weren't well understood until the 1960s—you might put a firm slab of Jell-O in a saucepan, put the saucepan on a burner, and light the flame.

One has to wonder what went on in the minds of southern California's growth lobby as they set out to make Los Angeles the largest city on the planet. (If that was not their stated intention, they acted as if it was.) The semi-arid basin that they were conjuring into the West's first megalopolis, in the late nineteenth century, had enough water flowing in its rivers, and accumulating in its aquifers, to grow at the rate it was growing for perhaps twenty more years. Then, had all irrigation farming ceased, growth could have continued for another couple of decades. At that point one would have had to ban irrigation on lawns, and in parks. With today's population, even with all outdoor water use banned—no swimming pools—the basin's local sources might permit everyone a bath once a month—in wet years.

The boosters may have believed the chamber of commerce "experts," who consistently overestimated the sustainable water supply by a factor of three or four. Or, more likely, they were deaf to caution, blinded by their own tub-thumping avarice. If Los Angeles ran out of water, it would find more—somewhere.

The options, if even imaginable, were drastically limited. Topography was Los Angeles' implacable enemy. Completely isolated by ramrod ranges and vast, austere deserts to the north and east and south, it had three rivers within two or three hundred miles to choose from (and no city in history had ever gone that far to augment its limited water supply). One was the Colorado, emerging from the Grand Canyon on the lee side of the Mojave and Sonoran Deserts. Then there was the Kern, flowing out of the Sierra Nevada into a great marshy lake—Tulare Lake, which has since disappeared. Finally, there was the Owens River, the largest of several streams draining snowfed watersheds along the eastern slope of the Sierra Nevada.

Fred Eaton was the first southern Californian who sensed exactly how little water the basin really had. He also knew where it would have to go for more. Eaton's forebears had been early pioneers, and he had been superintendent of the privately run Los Angeles City Water Company before serving a term as mayor from 1899 to 1901. His interest in water grew out of a quintessentially southern Californian mixture of paternalism and greed; he knew that his beloved city would inevitably run out, and would pay almost anything for more water that he hoped, in some sense, to own. In the 1890s, Eaton spent several vacations roaming southern California's watersheds, surveying and measuring streamflows. His explorations had shown him that if incoming new water had to battle gravity, it would never arrive. Hydroelectricity was still expensive and scarce, and growth was swallowing the region's oil production. Those realities ruled out any man-made river that had to be lifted over high elevations in order to reach the basin; they most emphatically ruled out the Kern River and the Colorado. Kern River water would have to surmount the Tehachapi Range, a four-thousand-foot summit to the north; Colorado River water would have to labor over San Gorgonio Pass, which was roughly as high. The alternative was tunneling. The European railroads had advanced the art of tunneling, but even the Swiss hadn't bored dozens of miles through a mountain range. Tunneling was also expensive and risky, especially in the sorts of fracture zones that surrounded Los Angeles.

The Owens River, on the other hand, was miraculously situated. It flowed through Owens Valley, a narrow finger of the high Mojave Desert between the abrupt escarpment of the eastern Sierra and the White Mountains to the east. Like all Great Basin rivers, it had no outlet; it terminated in a big lake. Owens Lake was 3,500 feet higher than Los Angeles, and no overweening topography stood between source and city. Some tunneling and siphoning would be required, but gravity could move the water all the way to the coast; at some steep elevation drops, it could even generate hydroelectricity to pay back the cost of construction. Eaton had looked at every alternative, and there were no others. Los Angeles had to capture the Owens River, or its growth had to stop.

On a fall day in 1904, Fred Eaton and William Mulholland climbed into a buckboard wagon and rode across the

Mojave (hardly anyone had done it before) to look at the river. Eaton had already seen it, at high flow, and came back obsessed. It was Mulholland who needed convincing. He was a rough but literate Irish immigrant whom Eaton had groomed as his successor at the city water company, even though Mulholland had no formal training as an engineer—a lapse in learning that would someday prove as fateful as their trip.

Mulholland returned impressed but uncommitted. The distance to the Owens River was intimidating—he and Eaton had used up hogsheads of water and many bottles of whiskey as they creaked along for almost two weeks. Most of the river's flow was already appropriated by valley farmers, who had created one of the few thriving irrigated meccas in the arid West. The Reclamation Service, the new federal agency whose mission was to irrigate, or reclaim, arid terrain—a task at which most private entrepreneurs had failed—wanted to build on the valley's success; it was sketching plans for a dam that would capture the river's flood flows, which could bring thousands more acres into reliable production. Los Angeles still didn't approach New York City in size or wealth, and in Mulholland's mind the prize seemed beyond his grasp. Not only would Los Angeles have to buy up a whole river (or steal it), but it would have to shove the federal government aside and then build an aqueduct system far grander than New York's—the greatest in human history.

That is exactly what it did. By 1904, Mulholland was finally persuaded that he had no options other than, as he sometimes suggested, to murder everyone connected with the chamber of commerce. In early 1905, Eaton, who had already purchased options on some water rights, returned to the valley to buy more. Whose money he used is not exactly clear, but most everyone who counted in Los Angeles—the newspapers, the politicians and oligarchs—knew of the plot. (To help ensure success, the local newspapers swore an oath of silence, which Harrison Gray Otis would break near the end, infuriating his rivals.) Eaton's great good fortune was his friendship with Joseph Lippincott, the regional director of the Reclamation Service, and Theodore Roosevelt's warrior interest in populous cities at America's western flank. Eaton persuaded Lippincott to hire him to settle a minor water rights issue; it was a bogus cover that gave him (probably illegal) access to deeds and records in Inyo County's files that he might need to cut the necessary deals. Then Eaton—posing as a wealthy rancher who hoped to amass a big spread in the valley—began optioning water rights. Because he had learned the financial situation of many farmers and mutual water companies, his purchasing instincts were dead-on; almost everyone he approached was willing to sell, especially at the prices he offered. Since no one suspected that his water was headed for Los Angeles, the farmers sold their rights without remorse.

The water, in so many words, was stolen fair and square, with the help of the federal government. Roosevelt's Reclamation Service abruptly bowed out of the valley. Lippincott resigned and went straight to work as Mulholland's well-paid deputy. Meanwhile, the U.S. Forest Service, under chief forester and close Roosevelt friend Gifford Pinchot, annexed a large swath of treeless Mojave Desert to the Inyo National Forest in order to clear a cheap, problem-free right-of-way for the aqueduct. Then, after the Owens River project was completed in 1913—Mulholland and the newspapers whooped the voters into such a drought panic that they approved the bond measure by a margin of fourteen to one—most of its flow never reached Los Angeles. It was diverted the moment it entered the basin and spread over the desolate San Fernando Valley, where a land syndicate comprising Otis, Chandler, the railroad magnate Henry Huntington, and other members of the city's capitalist royalty had recently bought a tract about five times the size of Manhattan. (They acquired the holding in 1902; whether that was early enough to know of the city's intentions is debatable, but in California, on the issue of water, the ravings of conspiracy buffs are too often true.) Owens River water that would, for many years, remain surplus to the city's needs let the syndicate cover their land with orange trees and other lucrative export crops. Back in the Owens Valley, roughly the same acreage, having lost its water, reverted to desert, and competition from the warmer San Fernando Valley crushed those farmers who held back some water rights and tried to hang on. In the end, San Fernando orange blossoms metamorphosed into tract homes, and the syndicate walked away with a profit that has been estimated in the hundreds of millions of World War II–era dollars.

It was a humiliation that inspired Owens Valley farmers to resort to dynamite. In the 1920s, sections of the 223-mile aqueduct were blown sky-high, again and again, until the city sent in trainloads of detectives and declared the whole region under (probably illegal) martial law. The episode earned Los Angeles eternal ill will in the rural West; most Wyoming ranchers know this history. But it did not end Los Angeles' quest for water; that had just begun.

Before Mulholland's career was buried, in March of 1928, under the ruins of his wretchedly mislocated Saint Francis Dam, whose collapse drowned hundreds of people, he had already concluded that the city must next reach for the Colorado River. It was no longer out of the question, as long as the federal government (the city's eternal rescuer) agreed to build the world's highest hydroelectric dam to power billions of tons of water up the intimidating quarter-mile lift. The dam—it was first called Boulder; now it is Hoover—was completed in 1936, after years of legal and political warfare between California and its neighboring Colorado basin states. A few years later, the Metropolitan Water District of Southern California, the region's newly formed water imperium, constructed another two-hundred-mile aqueduct to bring in water for another thirty years of growth.

Los Angeles was trapped in a vortex of excess. Its ceaseless promotion of itself had led to a population influx, which led to a water crisis, which led to the Owens River Aqueduct, which led to a greater population influx, which led to a new water crisis, which led to the Colorado River Aqueduct, which was about to lead to more people and another water crisis.

The city's tactics were those of a junkyard dog. By the 1950s, the Metropolitan Water District was taking twice its 550,000-acre-foot entitlement from the Colorado River. Its strategy was simple: it "borrowed" much of Arizona's unused entitlement, which was unused because California's congressional delegation, ten times the size of Arizona's, sabotaged any legislation that tried to authorize the project that would let Arizona use it. At the same time, the city extended the Owens River Aqueduct northward, across the divide leading to the Mono Lake basin, and began to divert most of the streams flowing into Mono Lake, a weirdly beautiful desert apparition where—strange as it seems—most of California's seagull population breeds. The lake level was to drop by dozens of feet over the years, concentrating salinity, and causing its once astonishing wildlife to decline. The Los Angeles Department of Water and Power began hiring battalions of lawyers who did nothing but defend it against environmental lawsuits.

It still wasn't enough. Mulholland's successors grew desperate; their plans became fantastic. One long-standing chairman of the Metropolitan Water District, Joseph Jensen, sermonized ceaselessly for an aqueduct from the Columbia River, a thousand miles away. The Department of Water and Power, whose ideas rivaled those of Metropolitan, purchased a piece of strategic canyonland on the Eel River, six hundred miles to the north, which contained an excellent dam site. (The dam was never built because Governor Ronald Reagan, who fought Indians in Hollywood, vetoed it; the reservoir would have submerged the Round Valley tribal reservation.) The North American Water and Power Alliance—NAWAPA—a hallucinatory scheme that would shanghai Alaskan rivers to southern California and Mexico, was conceived by Donald McCord Baker, an engineer with the Department of Water and Power. It was promoted full tilt by the same types who were promoting more growth.

The next jolt of water was to come, finally, from northern California's Feather River, via the longest aqueduct and the highest dam and the most imponderable pump lift ever engineered. The assessed valuation of greater Los Angeles guaranteed the bonds that financed the project's stratospheric cost—about $15 billion in 1999-value dollars—but the project was built mainly because it was the personal obsession of Reagan's predecessor, Edmund G. Brown. Pat Brown was one of the last of the New Deal Democrats; he "loved building things," he once confessed in an interview, and he was "absolutely determined . . . to build that goddamned water project—I wanted this to be a monument to *me*." (The governor, who grew up in San Francisco, had a disarming reason for wanting to send northern California's water south: "I don't want all of those people moving to northern California.") As Brown pushed his project forward, however, the political landscape fractured treacherously in his path. Northern California was violently opposed—more so than expected—but so, initially, was the Metropolitan Water District, which wanted to protect its ability to siphon away Arizona's unused Colorado River entitlement. The very idea that there might be another source besides the Colorado threw Metropolitan's solons into a fright, because it might undermine California's case against Arizona in the U.S. Supreme Court. (California lost anyway.) Meanwhile, the midwestern burghers whom the chamber of commerce kept seducing turned more conservative the longer they stayed; their ceaseless migration gave the project its raison d'être, but many voted against it because it would cost too much.

The 770-foot-high dam, the 444-mile-long aqueduct, and the 3,000-foot pump lift were ultimately approved by referendum and built during the 1960s and 1970s. The project has never delivered—and may never deliver—all the water it promised, but it won southern California another thirty-year reprieve. Where the region will go for more water is an interesting question. The cost of going to California's north coast, where undammed rivers still flow, is outlandish; besides, most of these rivers now enjoy Wild and Scenic protection, and opposition to new river dams is so ferocious that none has been built in the state since 1978. For now, it is storing surplus water, when it can find any, in offstream reservoirs, storing more in depleted aquifers, and paying farmers to use less. Purified seawater or hauled-in icebergs may be its ultimate fallback. But for now, at least, the miracle persists: fifteen million people inhabit a region with enough sustainable water for a population ten times smaller, and, every day, two billion gallons flow across deserts and climb or flow through mountains so that they may remain. Half of it goes to golf courses, pools, cemeteries, parks, and lawns.

Many people know some of this history; most still don't fathom that the Bay Area did much the same thing. As Owens Valley farmers were dynamiting Los Angeles' aqueduct, San Francisco was hooking a catheter into its own veins, leading water from a reservoir behind the only big dam ever built inside a national park.

The dam, named after an engineer, is called O'Shaughnessy; the valley it inundated, named by a Sierra tribe, is called Hetch Hetchy; the man who died fighting its inundation was John Muir. The city could have siphoned rivers that cut through less spectacular valleys, but it chose the Tuolumne River and the Hetch Hetchy site for the same reasons Los Angeles went to the Owens River: engineering elegance, abundant runoff, and impressive hydroelectric potential. The water diverted from Pilarcitos Creek had sufficed for half a century; now it was no longer enough. Across the bay, Oakland and a consortium of neighbor cities faced the same dilemma in the burgeoning twenties and thirties, and solved it the same way—with heroic engineering and no apologies.

Except for a few hundred thousand people who are supplied by local sources (my own county, Marin, dammed its own small streams), the population of the Bay Area depends on these two great lifelines—a 155-mile aqueduct leading from the Tuolumne River, and another 90-mile aqueduct from its neighbor watershed, the Mokelumne River—and on branch aqueducts from the federal and state water systems. By and large, our attitude toward these giant engineering feats is a collective shrug. But, like everyone in Los Angeles, we who live in the Bay Area are hooked on imported water. Also, at the risk of californicating the simile, we are in denial. We never imagine that, someday, the water might not arrive.

II

The plate tectonics revolution, which convincingly explained the dynamics behind earthquakes and gave geologists some capacity to anticipate them, began to gather force in the twenties. Unlike the concurrent revolution in submolecular physics, it did not move apace. When the Beatles made their appearance on the *Ed Sullivan Show,* plate tectonics theory still qualified as heresy. Even after many faults had been traced and mapped, and a spectacular concentration of faults was evident around the Pacific Rim, earthquakes were widely regarded as random events, response mechanisms to uneven cooling and hardening in the earth's interior. Had geologists recognized a century earlier that the continents are adrift, riding transient segments of the earth's crust that bump and crush against one another, causing episodic earthquakes in the same general area—had anyone understood this, then California might have been settled differently. Its population might have gone inland, away from the most hazardous seismic zones. Fewer people might have emigrated here.

More likely, though, everyone would have come and settled exactly where they did—just as, in Florida (another state experiencing insane growth), most people populated the most hurricane-haunted of the two coasts. Sun, sea, warmth, jobs, solace in old age—these seductions conquer fear and memory. In the late 1890s, 80 percent of California's population had settled in regions where the preponderance of its major earthquakes have occurred. In the late 1990s, that figure was unchanged. It won't change, at least not much. What's there is there.

After discovering plate tectonics, geologists recognized that earthquakes occur in regions astride fault boundaries, where huge slabs of the planet's crust, moving contrarily, lock together and accumulate strain. As the strain becomes intolerable, sequences of faulting relieve it. Sequences tend to run into decades: periods of relative quiet yield to periods of heightened—or intense—seismic activity. Reeling in the centuries from historical records in China and Japan, one is struck by several eras when earthquakes were suddenly more numerous and powerful than before or afterward. One such period in Japan stretches from the late 1700s to the early 1900s; Tokyo, then called Edo, was destroyed three times. California's reliable written record is only two hundred years old.

The first notable earthquake mentioned in mission-era records was in 1800; the first deadly one hit in 1812, a year when several quakes were felt—the priests later called it *el año de los temblores*. The epicenter of the 1812 quake may have been near Ventura, north of Los Angeles, but the only inhabitants were the native people, who lived in reed-and-mud huts. The worst structural damage was at San Juan Capistrano, sixty miles south. A Tuesday mass had packed the mission with worshipers, and thirty-four of them died when first the roof and then the walls collapsed.

The next major event occurred in 1836, in the Bay Area. For decades seismologists believed that the fault responsible was the Hayward, an offspring of the Mother of California faults, the San Andreas. Lately, however, some have speculated that the epicenter was on a length of the San Andreas fault near Santa Cruz, or perhaps on another of the Bay Area's known (or unknown) faults. Anecdotal reports aren't much help, because there weren't many and almost none have survived; San Francisco's population was eight hundred or so, and across the bay there were more bears than people. Nothing much had changed when the next powerful earthquake struck two years later. The 1838 quake was probably precipitated by the peninsula segment of the San Andreas fault, between San Jose and San Francisco. Its magnitude on the Richter scale may have been a tick or two under 7.0, much like the temblor two years earlier. What's significant about these two earthquakes is that their recurrence today, two years apart, in a region of seven million people, is unthinkable.

The 1836 and 1838 quakes were an anomaly in their era, which was otherwise quiet. During the first eight decades after Junipero Serra arrived, there were only five earthquakes in colonized California with an inferred Richter magnitude of 6.5 or greater. Then, during the next five and a half decades, there were at least *eighteen* earthquakes whose magnitude was greater than 6.5.

The second in the cluster, on January 9, 1857, was the earthquake of the century. The epicenter was near Tejon Pass, in the Tehachapi Mountains, where the San Andreas fault and a giant transverse offspring, the Garlock fault, conjoin along the present-day section of Interstate 5 known as the Grapevine. The pueblo of Los Angeles was sixty miles away. The nearest settlement was a forlorn army outpost called Fort Tejon, of which a few remnants survive. An officer stationed there watched one side of a nearby mountain separate and rumble downslope as a forest-clearing landslide. The quartermaster, who wrote the official report, described "the most terrific shock imaginable, tearing the officers' quarters to pieces, severely damaging the Hospital, and laying flat with the ground the gable ends of the buildings. Immense trees have been snapped off close to the ground." Sand blows—geysers of hyperpressurized groundwater mixed with soil—blew out everywhere. A rift line still discernible in places corroborates a startling thirty feet of horizontal ground displacement—ten feet more than at Point Reyes in 1906.

Depending on whose hindsight you believe, the Fort Tejon earthquake was a 7.9, an 8.2, possibly even a 8.3 on the conventional Richter scale. It was huge, a monstrous release of arrested energy—many hydrogen bombs' worth—and it created phantasms across half the state. At the southern end of the Great Central Valley, near the base of the Tehachapi Range, what became a palatinate of industrial agriculture was then a shallow, evanescent lake called Tulare. It was the terminus of four Sierra Nevada rivers, the Kings, Kaweah, Tule, and the Kern—four rivers of such recent origin they hadn't yet eroded an outlet to the sea. They pooled at the bottom of the valley, and in drought years the lake was dry by midsummer. During big snowmelt years, the lake's surface area in spring was three times the size of Lake Tahoe's. On the day of the Fort Tejon quake, a party of ranchers riding horses near the shore was transfixed when millions of waterfowl erupted simultaneously over the lake's surface. Then, one of the travelers wrote, "The lake commenced to roar like the ocean in a storm." A long series of crashing waves thundered toward shore, panicking the horses, which galloped off the other way. When their riders coaxed them back, they found fish flopping three miles from shore. "They could have been gathered by the wagonload," one of them observed.

Gold miners along the Kern River, which was closest to the epicenter, said that the river momentarily stopped flowing. Then it flowed backwards. Then, volcanically, it spilled over its banks. The Mokelumne River was a hundred miles north of the Kern but put on a better show. Hundreds of miners panning gravel along its banks watched the river jump out of its bed, exposing its glistening rocks, then cannonball back in and go on its way, just as before, except where landslides had caused new rapids to form.

In Los Angeles, most buildings were single-story and constructed of adobe, which is vulnerable to seismic shaking. But the pueblo sustained relatively little damage; why is still a matter of debate. Either most of the energy was broadcast in another direction, or else the buildings' thick walls and low heights helped them stand up to the shock.

In 1865, eight years after the Fort Tejon quake, there was a powerful temblor south of San Francisco, again on the San Andreas fault. (It was the same section of fault that would rupture once again, in 1989.) Trees on the peninsula were severed in half, and San Francisco—which was growing out into the bay—sustained appreciable damage in the parts of town recently erected on fill. One witness to the damage was Mark Twain, who was back in town haggling with the *San Francisco Call* over rights to a series of travel stories destined to become *The Innocents Abroad*. Walking along Third Street, Twain watched a rolling seismic wave open a fissure in the street, which panicked some horses pulling a streetcar, and their fright threw a passenger halfway through a glass window, where he lay impaled.

Then, in 1868, the earthquake that would reign for almost four decades as the Big One struck.

Twenty years had come and gone since James Marshall glimpsed gold in the American River. San Francisco was the largest city west of the Rockies, and Oakland was not too far behind. New villages were growing up around the bay—Hayward, for example, a South Bay town founded amid orchards and pastures feeding the bigger cities to the north. On Wednesday morning, October 21, at six minutes before eight o'clock, the destruction of Hayward began. That was when the first jolt was felt. According to the *History of Alameda County Under the Stars and Stripes*, published in 1876, a series of aftershocks arrived all day long, taking the town down serially:

> First shock, at 7:54 A.M., very heavy—direction, northeast, east and southwest, a rolling motion. Almost like a continuation of this came a whirling motion. 8:26, slight shock; 8:44, heavy shock, with a rolling motion; 9:11, slight shock; 10:15, heavy shock, with rolling motion, and up and down movement; 3:12, slight shock; 3:17, slight shock; 4:08, double shock, up and down.

There were shocks during the entire day; some observers stating that there were thirty-two. Those we have named attracted attention and were noted.

Oakland, fifteen or twenty miles from the epicenter, was surprisingly undamaged—miraculously is more appropriate. According to the *History*, "A stranger passing through its streets immediately after would not suppose that anything unusual had happened." The worst damage was, predictably, at the waterfront. At Alameda, built on a low island in the bay, "scarcely a house escaped uninjured." There was a lot of damage in low-lying parts of San Francisco, where collapsing buildings took several lives. Around San Jose, "Great damage was sustained by many buildings. The large and elegant Presbyterian Church, with its tall tower, was very badly injured. The tower was so badly cracked that it had to be taken down . . . [and] the organ was destroyed by bricks falling through the roof." Every building in San Leandro, the next town south of Oakland, was reportedly damaged or destroyed. "Bad as is the destruction of property in San Leandro," goes the account, "it is worse at Haywards"—as it was then called. "[M]any wooden buildings fell which would have remained secure if they had been built upon proper foundations. . . . The flouring mill owned by Sheriff Morse is a complete wreck. . . . The immense grain warehouse of Mr. Edmonston is as completely ruined as it could possibly be. . . . Washington Hotel [is a] complete ruin." At Martinez, at least forty miles from the epicenter, a large warehouse was seriously damaged, and many fissures opened in the earth. Springs began to gush all over the place, and a dry creek through San Leandro was suddenly engorged with three feet of flow. There are no accounts of significant fires in the aftermath, and that was dumb luck. The first rain of the season fell almost a month later, on the evening of November 18.

Finding opportunity in the calamity, Hayward, several decades later, would build a new city hall next to a leafy glade nourished by one of the earthquake's newly formed springs. For now, the building is still there, but has been condemned: seismic activity along the fault is continuing to weaken the structure.

The 1865 and 1868 earthquakes were a reprise of the two closely spaced large earthquakes that struck the Bay Area in 1836 and 1838. One lesson from these two serial events was that, for the sake of caution, the region's towns and cities might best relocate. Another lesson—since masonry was replacing wood as a building material—was that wooden structures withstand ground motion better than masonry, so long as they are bolted to their foundations. A third lesson was that the most vulnerable structures sit on loosely consolidated soils—especially semi-saturated landfill—and if the region was going to grow, at least expansion into the bay should stop.

Four years later, California was hammered by another earthquake—a quake that, by contemporary definition, straddled the divide between major and great. Every notable earthquake since 1800 had struck along or near the coast, presumably caused by plate movement along the San Andreas or one of its proximate faults. Yet the epicenter of the 1872 quake was far inland. It was in the Owens Valley, two hundred miles east of Santa Barbara, and was caused by strike-slip faulting. (In strike-slip, one great block of landscape lurches horizontally—in California, that usually means northwestward—while the adjacent one stands still, more or less.) A long cliff-like scrap on the Western side of the Owens Valley remains as a testament that, in a few seconds in 1872, several miles of terrain underwent up to thirty feet of uplift and horizontal displacement. In Lone Pine, the valley's largest settlement, forty-three of the fifty structures survived. On the Modified Mercalli scale, which rates historic earthquakes—those predating seismometers and the Richter scale—based on observations of structural damage, visibly displaced terrain, and observed phenomena like levitated rivers, the Owens Valley earthquake caused Level X intensity. It was one of the three largest California earthquakes since 1800. The epicenter was near the future site of a very large dam.

In 1873, there was an earthquake, retrospectively rated plus or minus 7.0, near the Oregon border. In 1892, a 7.0 near the opposite border with Mexico. A string of 6-pluses through that decade: three in 1892, another in 1897, two more in 1898, another two in 1899. The apotheosis of the after-1857 swarm came in 1906, the first North American earthquake whose epicenter was on the outskirts of a major city. The San Francisco earthquake remains the most destructive in U.S. history and, in the Western Hemisphere at least, the most famous earthquake that ever was. Compared to other earthquakes, however, fatalities in the San Francisco quake were low. At most, three thousand people died; in 1976, the Tangshan earthquake killed as many as half a million Chinese, and most people have never heard of it.

Among famous earthquakes, 1906 may be the best known and the most misunderstood. The city of Santa Rosa, fifty miles north of San Francisco, sustained much worse damage, proportionate to its size, from ground motion; it was virtually leveled. Santa Rosa sits on poorly consolidated alluvial soil. San Francisco—more of it then than now—was built on rock. The Franciscan Assemblage's rock is granite's antithesis—you can pull chunks out of cliffsides with your bare hands—but, like most bedrock, it conducts seismic energy well. In 1906, the most extensive damage caused by shaking was in the same areas that were badly damaged in 1868: the low-lying parts of the city, especially those erected over fill. The bedrock city, the city sprawling over its hills, remained largely intact for a day and a half, until the holocaust.

That fire, not violent ground-shaking, was responsible for most of the city's destruction in 1906 is a fact that San Francisco's political leaders took pains to camouflage. They had their reasons. An earthquake was, after all, an act of God. One couldn't do much about earthquakes, except build stronger structures. On the other hand, since the earliest days of the mission era there had never been an earthquake this powerful in northern California. The next one like it might be centuries off.

Fire, on the other hand, was a force one could fight, as long as one was adequately prepared. Just seven months before the 1906 quake, the National Board of Fire Underwriters had declared San Francisco's thirty-six-million-gallon-per-day water system inadequate in the event of any large conflagration. "San Francisco has violated all underwriting traditions and precedents by not burning up," the report said. "That it has not done so is largely due to the vigilance of the fire department, which cannot be relied upon indefinitely to stave off the inevitable." The city's fire chief, Dennis Sullivan, who was to die in the earthquake, knew all this; for years, he had implored the board of supervisors to appropriate more money to beef up the system. But the supervisors, many of them protégés of Abraham Ruef—the famously corrupt kingmaker who ran the city on behalf of its pliable musician-mayor, Eugene Schmitz—seem to have had other fish to fry, or nests to feather.

Even with three times the water capacity at hand, much of San Francisco would have burned anyway; amid the chaos and debris and steep ruined streets, horse- and cable-pulled equipment couldn't get to the many small blazes before they joined and became an invincible force. That, of course, was the city fathers' gnawing fear. A city helpless against fire was a city apt to scare off potential investors and insurers. Outside of downtown, which was built of masonry and brick, San Francisco's building stock was constructed mainly of wood—a wooden city in an earthquake zone with summer-long droughts. San Francisco was, and may still be, more susceptible to conflagration than any other great city in the United States. It was a fact no one wanted to advertise.

The conflagration began as isolated blazes, ignited by unattended stoves, leaking gas on sparks, arcing voltage, volatile chemicals—the same causes apt to start a similar fire today. After wrecked pipes and mains had emptied most of the water system, big, explosive fires formed and were anointed with names; the biggest was the Ham and Eggs Fire. In the end, they converged into a two-pronged monster that engulfed all of downtown, ran through most of the Mission District, and swept like a rotating scythe over Nob, Russian, and Telegraph Hills. (On the first day prevailing winds blew the fires west, toward the ocean; then an ocean wind blew them back east into neighborhoods that had up till then been spared.) One of the fire fronts burned itself out at the bay after devouring the city's entire Chinese and Italian communities; the other was stopped at Van Ness Avenue, where the army had dynamited all the elegant buildings on both sides of the broad boulevard. By then the fire had swept through 520 square blocks and incinerated close to 30,000 structures, including most of the city's great mansions. Even ships were set afire by the embers and destroyed. Two hundred fifty thousand people were instantly homeless, bereft of savings (which burned up in banks), without water, and with almost nothing to eat. The entire economy was wrecked. For months, most San Franciscans lived, slept, and cooked outdoors. What had been ruined in three days would take a decade and longer to rebuild. The cluster of strong earthquakes in the late nineteenth century had ushered out the era of whale oil and wood. Abetted by rapid population growth, 1906 showed what was in store in the dawning age of oil, gas, and electricity.

Six years after the great San Francisco earthquake and fire, Alfred Wegener, a modestly prominent German meteorologist, offered his theory of plate movement, or continental drift, in Marburg University's *Journal of Science*. In 1915 he published a longer version as a book, *The Origin of Continents and Oceans*. The lordly title was unfortunate, since Wegener's hypothesis contravened what nearly all of the geologic profession believed, and worse, the bedrock convictions of that conservative establishment were being challenged by . . . a *weatherman*. "This book makes an immediate appeal to physicists," wrote a reviewer in *Nature*, then the world's leading scientific journal, "but is meeting with strong opposition from a good many geologists."

That was understatement. The reaction from the geologic profession amounted to whoops and roars, howls and knee slaps, Bronx cheers and whistling salutes. One famous geologist, delivering a venomous lecture on Wegener's apostasy, held up two halves of a trilobite, one from Europe and the other from the Americas, and said the animal must have been torn in half when the continents separated.

Some of Wegener's calculations turned out to be wildly flawed—he had Greenland migrating at a rate of kilometers per year—and his mistakes were fodder for his enemies. But the hypothesis was irresistible. Nonetheless, the geologic establishment resisted as long as it could, causing Wegener such consternation that he died a miserable man. (And Wegener wasn't even the first to hypothesize continental drift; the basic theory was put forth a century earlier by a Scottish philosopher named Thomas Dick.) Wegener finally became a prophet, his vindication posthumously complete, at the annual conference of the American Geophysical Union in 1967, just a couple of years before the first Apollo moon landing. Our current understanding of macrogeology, of the immense forces behind the shape of the visible world, begins there.

Plate tectonics unveiled phenomenal mysteries: the dynamics of mountain building, of seafloor spreading, of vulcanism and the accretion of exotic terrain. It explained why Point Reyes, the peninsula north of San Francisco, is a granitic stranger in a milieu of crumbly Franciscan rock: the whole landform migrated up from Mexico on the Pacific Plate. It explained the volcanic peaks of the Cascade Range, forced upward by subduction faulting at the interface of the North American, Juan de Fuca, and Gorda Plates.

What the theory could not explain, or at least predict, was the frequency of earthquakes on a given section of fault. The answer to that riddle had to wait a few more years. The work was performed by a young Caltech geologist named Kerry Sieh, pursuing earlier efforts by Bob Wallace, Lloyd Cluff, and Chuck Taylor. Sieh's inspiration, brilliant in its simplicity, was to trench into the streambeds, isolate the displaced strata, and carbon-date sediments from each dislocated layer of ancestral stream bottom. Sediments from six feet down might be dated back to, let us say, 1100 B.C. An overlying layer, pushed a few feet to the north by displacement during a major earthquake, might be dated back to 920 B.C. More recent stream displacements may have occurred at intervals of two, three, or four hundred years. With three thousand years' worth of carbon-dated sediments deposited in discrete cross sections of migrating streambed, you have enough data to calculate the average historic interval between earthquakes. If you know when the last one occurred—1857 on the trans-Tehachapi section of the southern San Andreas fault; 1868 on the southern section of the Hayward fault—you have a rough idea when the next may strike.

Streambed excavation, or trenching, has become the bread-and-butter work of a number of geotechnical consultants. One of them, in the Bay Area, is the firm of William Lettis & Associates. Between 1995 and 1997, Bill Lettis, a handsome and soft-spoken man in his mid-forties, endured several marathon interviews with me. A couple of times, between sessions, he invited me to trenching parties, inaugurating a new ditch at some new focal point on a fault—the Calaveras, the Rodgers Creek, the discontinuous Sierra Foothills fault zone. Any time spent with Lettis at his office ended with us half-buried in maps, in cross-sectional analyses, in GATS imagery taken from satellites and U-2 aircraft. I remember how, early on, this stuff was utterly opaque to me. Gradually, though, it attained fearsome clarity.

"You might consider this an inexact science or art," Lettis once told me. "Geologists think in terms of millennia, eons, ages. So if we're correct to the nearest century we think we're being exact. In seismology, that's about as close as you can come now. A lot of earthquake recurrence we can't even guess at. Blind thrust faults, the kinds of faults responsible for [the] Northridge and the San Fernando earthquakes—you can't trench those faults. They're too deep. Until recently, we didn't recognize the earthquake potential of these faults and, in some instances, that these faults even existed. We now know that a large thrust fault underlies Mount Diablo and projects toward the surface beneath the rapidly urbanizing Livermore Valley. In the San Francisco Bay Area, however, the most important faults are the strike-slip faults, like the Hayward and San Andreas, because they are capable of producing M7 to M8 earthquakes every two hundred to four hundred years. We have been trenching these faults for years to determine when and how frequently past earthquakes have occurred. If we can establish the past pattern (i.e., size, location, and time) of past earthquakes, then we can make an educated forecast of the likelihood for the next earthquake on the fault. At this time, we think we have a pretty good idea what's gone on here historically. The short answer is that these faults go like greased lightning—as faults go. They move in frequent steps. Even in a huge earthquake like 1906, most of the San Andreas fault doesn't slip. It's that big—that long and complex. The static segments catch up later. The southern section of the Hayward fault ruptured in 1868 and caused the M7 Hayward earthquake—we're convinced of that. It's conceivable that a relatively short, sixty-mile fault like the Hayward could all go at once. There are gigantic forces to be overcome—friction and inertia involving billions of tons of rock. We have little quakes all the time, earthquakes only instruments can sense. They relieve some strain. On some faults, like the Hayward, we see a fair bit of creep. Creep is just slipperiness—movement that doesn't even qualify as a small earthquake.

"The problem is, a great earthquake like 1906 releases millions of times more energy than a tiny quake a few notches down the Richter scale. That's the paradox of a logarithmic scale. Sooner or later, there *has* to be a big earthquake. Depending on the fault length and the amount of stored strain, 'big' can mean a strong quake—a 6.0 to a 6.9—or a major quake—let's say between 7.0 and 7.7—or, on a long fault like the San Andreas, a great earthquake like 1857 or 1906, which were approximately 8. These are kind of arbitrary numbers. Sooner or later a magnitude 7 plus earthquake will strike the Bay Area. A 7.2 epicentered in the Bay Area is . . . well, you can't imagine the damage and loss of life it could cause. What does 'sooner or later' mean? All we can do is take the incidence of past known earthquakes, which is an average number, and use the faults' rate of movement—the slip rate, which is about one inch per year on the San Andreas fault—to estimate the likelihood of the next, and then bet your odds. In 1988 we convened a panel of geologists and seismologists to estimate the probability of an M7 plus earthquake occurring in the Bay Area in the next thirty years. Based on past earthquake activity, we divided the San Andreas fault into several distinct segments, each of which could produce an M7 plus earthquake. We identified the Peninsula Segment, extending from San Francisco to Los Gatos, and the Santa Cruz Mountains segment, extending from Los Gatos to San Juan Bautista. We were pretty sure that one or both of these segments broke in the 1865 earthquake. Based on our reading of the rock, we came up with roughly a 28 percent chance of a major earthquake within the next thirty years for these two segments. Those were the consensus numbers we used in our report. I'm sure a lot of people thought, 'Not to worry, nothing's gonna happen for at least thirty years.'

"Eleven months later, the Santa Cruz Mountains segment slipped, and produced the M7.1 Loma Prieta earthquake."

When Lettis was explaining this to me, we were sitting in his office in downtown Oakland, on the fourth floor of a twelve-story office building erected before the San Fernando earthquake of 1971 rewrote the California Uniform Building Code. A few weeks after this interview in 1995, a devastating earthquake pounded Kobe, Japan. Several downtown office buildings similar to Lettis's underwent mid-level collapse—the upper floors fell onto the lower floors, and the middle floors, acting as energy-absorbers, were squashed, along with any life-form larger than a mouse. Had the Kobe quake, which struck very early in the morning, waited another few hours, there would have been thousands more fatalities just because of mid-level collapse.

Three months later, when I spoke with Lettis again, he had moved into a brand-new, low-rise building in Walnut Creek, fifteen miles off the fault.

. . .

After 1906, California entered an era of seismic quietude. Until 1971, the most damaging earthquake in California was the 1933 Long Beach earthquake, a moderately strong 6.3 on the Newport-Inglewood fault, which runs from under Los Angeles International Airport south to Newport Beach. Two generations earlier, much of the terrain along the fault had been marshes and duck clubs; two generations after the quake, it had become humanity-in-a-can. The damage, although considerable around the city of Long Beach, was nothing compared to what it would be today. The strongest earthquake during this period, the M7.2 Kern County earthquake, struck in the southern San Joaquin Valley, which was not yet the empire of agribusiness it has become. Bakersfield was far enough away, or just lucky enough, to escape substantial harm. There were nine other temblors that achieved Richter readings of 6.5 or greater, but they were either offshore or at the edge of nowhere. In California, the middle of nowhere was going extinct.

Then, in 1971, the San Fernando quake hit.

San Fernando was the sort of earthquake Californians had almost trained themselves not to fear. It struck along a section of a blind thrust fault—blind because it was deep and unmapped—just twenty miles long. (In thrust faulting, one piece of planetary crust dives beneath, or thrusts itself over, the adjacent one.) The epicenter was at the northeastern edge of the San Fernando Valley, twenty miles from Burbank, in a suburbanizing corner of the Los Angeles Basin where you still had farms. It was not a Big One, nor even a Little Big One; it was a mere 6.4. But it caused destruction no one had imagined. Four hospitals were severely damaged. The Veterans Hospital at San Fernando was almost completely destroyed—most of the fatalities were there—

and the Olive View Hospital, which had just been designed and built under up-to-date seismic safety codes, had two of its buildings collapse. (One housed all the ambulances.) The intense shaking brought down overpasses on freeways that had recently been dedicated. The highest levels of ground shaking were measured four times greater than what the California Uniform Building Code had assumed possible during an earthquake. It whacked thirty feet off the crest of the Van Norman Dam, which was, according to engineers, about eight seconds from total collapse when the ground motion died out. (The reservoir behind the dam was larger than the one whose sudden release wiped Johnstown, Pennsylvania, off the earth in 1889.) Eighty thousand people lived in what would have been the flood's path. The earthquake also created phenomena that, in their way, were as startling as imploding hospitals and leveled overpasses and decapitated dams. Cars were bounced into the air and turned halfway around. Windows and their frames flew out of homes. A man eating breakfast at his backyard patio looked up in time to see most of the water in his kidney-shaped pool levitate into midair. The water was shaped like a kidney.

The San Fernando earthquake sent the state's, and a fair portion of the world's, builders and architects and engineers back to their stress equations and policy-planning groups. In 1973, the Uniform Building Code was revised so extensively that future insurance rates would be strongly based on whether a structure was erected before or after that year. As more resilient public buildings were erected, and older ones were retrofitted or torn down, more strong earthquakes began to strike California, much as in the final decades of the nineteenth century. Two big jolts hit in

1980—one off the northern coast and the other near Mammoth Lakes, in the eastern Sierra, where the earth had been giving off ominous signs. Three years later, a magnitude 6.5 earthquake destroyed downtown Coalinga, a small city in the most godforsaken reaches of the San Joaquin Valley. It occurred on an unknown fault which, geologists now believe, underlies much of the western side of the valley.

Then, in 1989, the M7.1 Loma Prieta earthquake—like the San Fernando in 1971—struck at the fringe of one of the state's most populous regions. The epicenter was in the Santa Cruz Mountains, seventy miles south of the Golden Gate Bridge. Loma Prieta claimed only sixty-five lives, but caused $7 billion in damage; it was the latest lesson in the risks of laying gargantuan infrastructure atop a migratory landscape. More than half of the casualties were people trapped in their cars on the Nimitz Freeway, Interstate 880, where it veered west of downtown Oakland as a double-decked and elevated viaduct. The viaduct, which was known as the Cypress structure, amounted to several tons of concrete and steel per linear inch, but was anchored in soils as weak as the structure seemed indestructible. Seismic energy that traveled quite harmlessly through bedrock close to the epicenter was greatly amplified when it plowed into the unconsolidated soils around the bay. The Cypress structure, according to witnesses, lurched with a ponderous, manic rhythm—sideways, back and forth, up and down—as the underlying ground lost cohesiveness. The upper deck crashed onto the lower deck, the supporting columns fractured—the sound was that of saturation bombing—and section by section the monstrosity boomed to earth. A mile and a half went down, to varying degrees. During the next weeks, as thousands of truckloads of concrete and mangled steel were hauled away (where did it all go?), psychologists offered counseling to workers who had pried out cars flattened into six-inch slabs of metal, rubber, plastic, and human gore. Had the second game of the 1989 World Series not been about to get under way—and had the competing teams not been the Oakland Athletics and the San Francisco Giants—ten times as many vehicles might have been traveling the Cypress structure during its moments of doom.

Five years later, in 1994, the Northridge earthquake, an M6.7, struck along a blind thrust fault a few miles from the blind thrust fault that caused the San Fernando quake in 1971. Again, many things that weren't supposed to happen, did. Freeway overpasses designed to withstand more powerful shaking collapsed. Portions of Interstate 10, the Santa Monica Freeway, the umpteen-lane Mother of California freeways, were shut down for a month—for *only* a month, thanks to a reconstruction effort that rivaled the Manhattan Project. (In San Francisco, after Loma Prieta, it took seven years to rebuild the damaged overpasses at the confluence of Route 101 and Interstate 280; the two regions take a different view of the importance of vehicular liberty.) Much of the building stock around Northridge had been erected after 1973, and almost everything postdated the 1933 Long Beach quake, which had also caused revisions to the Uniform Building Code. Nonetheless, a hundred thousand people were instantly homeless—their houses and apartments were either wrecked or unsafe to occupy. At Northridge State University, 107 buildings were damaged at a cost of $350 million; future classes would be held in a couple of hundred trailers and modular buildings hauled onto campus.

The quake's strongest shock affected the structural engineering fraternity. A number of commercial buildings—brand-new, state-of-the-art, steel-framed, stress-unloading, flexible mid-rises—were badly damaged in the quake. The damage was invisible, for the most part; it affected the beam-column joints where the structures' steel endoskeletons met. Carefully inspected welds were ripped apart, and giant bolts were shorn in half; the buildings still stood, but they stood like three-hundred-pound defensive tackles with mangled anterior cruciate ligaments. Years later, a number of the weakened buildings were still unrepaired; simply getting at the damage means ripping out the structures' insides, and it may be cheaper to tear them down and build anew. Flying into Burbank Airport after the Northridge quake, you skimmed over some of these ghost-ridden office towers, surrounded by parking lots bereft of cars. On the approach to Burbank, you have a good view of downtown Los Angeles off to the south; you could contemplate identical structural weakness in, say, a sixty-four-story high-rise. By 1997, as more $50-million buildings were being repaired or simply written off as losses, the economic damage caused by the Northridge earthquake was gaining on Hurricane Andrew, which reigned, tentatively, as the most expensive disaster in U.S. history. Before World War II, the Northridge earthquake would have caused modest damage; that end of the San Fernando Valley was mainly farms. Before World War I, damage would have been negligible. Now it was likely to become the country's most expensive peacetime disaster. But from a seismological perspective, the quake didn't amount to much.

III

The good news is that the southern segment of the Hayward fault and the northern section have apparently never ruptured at the same time. If the whole fault displaces or if the northern segment and the Rogers Creek fault rupture simultaneously, we'll have a substantially bigger quake than the one we expect. A 6.9 is plenty big.

—Richard Eisner, Regional Administrator in Oakland for the California Governor's Office of Emergency Services (OES)

The Nojima fault segment that ruptured under Kobe is shorter than the Hayward fault. About 50 kilometers of fault rupture occurred during the earthquake. The Hayward fault could produce fault rupture of 60 to 70 kilometers, or even up to 150 kilometers if it ruptures with the Rodgers Creek fault north of San Pablo Bay. We now give a low, but finite probability that rupture on the northern Hayward fault could trigger movement on the Rodgers Creek fault. The two faults are separated by about a 6-kilometer-wide stepover centered in San Pablo Bay, which is probably why the San Pablo Bay depression is there. We estimate that the next earthquake on the Hayward fault could have seven to ten feet of horizontal displacement on about 5 to 10 percent of the fault trace, with an average displacement of about 3 feet on most of the fault. During the 1906 earthquake, average displacement was about twelve to fourteen feet north of the Golden Gate, and reached a maximum at Point Reyes, where close to twenty feet of surface rupture occurred. At Kobe, despite all the damage, only about three feet of surface rupture occurred. In the San Francisco Bay Area, that kind of surface displacement will be common. Poor soils are always a huge factor—that's one of the main reasons Kobe was so badly damaged. The soft soils and artificial fill placed around the Bay margin produced amplified strong ground shaking and extensive areas of solid liquefaction. We're going to have much more extreme liquefaction during a Hayward fault earthquake than we had in Loma Prieta. Conditions here are strikingly similar to Kobe, and we could have a stronger earthquake.

—William Lettis, of William Lettis & Associates

Sixty-two to 75 percent of our customers will be without water for several days if none of EMBUD's seismic retrofits are completed. In that event, it will take several months to get everyone back on line.

—David Pratt, Manager of Design
with the East Bay Municipal Utility District

The bottom line is that we're going to have a very difficult time moving around. The BART tunnel through the hills to Concord could be out of service indefinitely. The Caldecott Tunnel on Route 24 is likely to sustain major damage. On the west side of the tunnel, the highway will be sheared by ground displacement. The Warren Freeway will be undrivable for a long time for the same reasons. Sections of freeway built over rock or good soil conditions can be made passable fairly soon, even if they're knocked out of whack at the fault intersection. But liquefaction along Interstates 80, 580 to the north, and 880 to the south could pose formidable repair problems. We'll also see liquefaction along Route 101 in Marin County, at Corte Madera. The Oakland Airport will see major liquefaction damage to runways and may be closed for months. At San Francisco International Airport, damage will probably be similar to 1989. It should be up and running again in days or weeks. BART will be damaged around the West Oakland station, too, where the elevated track is likely to slump due to liquefaction.

—John Egan, Principal Engineer with Geomatrix Consultants

The three I'd most expect to see knocked out are the Bay Bridge—especially the Bay Bridge—and the Richmond– San Rafael and Dumbarton bridges. If we retrofit them in time, obviously things will be different. The question is, where's the money to do it? We're still going to have a hell of a time getting *to* the bridges. Ground motion and lateral spreading at all the eastern approaches will be extreme. The pavement will be a mess. Until they regrade the damage, you'll need a four-wheel-drive vehicle to get through. A lot of overpasses will be down or dangerous. The highways around the north and south ends of the bays are all on poor soils and are likely to be out. So it's gonna be, "Can't go over, can't go under, can't go around." The state has been subsidizing private ferries just to keep some around for this emergency. If you have a boat and a mountain bike, you can probably navigate in passable local areas.

—Ed Bortugno, senior geologist with OES in Sacramento

Northridge has opened our eyes to the vulnerability of high-rise buildings. They're vulnerable. They're vulnerable! We never expected modern steel-frame high-rise buildings to be vulnerable until we saw the joint stress and bolt shearing after Northridge. An earthquake always brings surprises. They're not always bad surprises. We've been surprisingly lucky with the most recent earthquakes. Kobe happened at five in the morning, before downtown was mobbed and the freeways were full. Northridge was at four in the morning. Loma Prieta—everyone was watching the World Series. Freeway casualties, especially on the Nimitz, could have been a lot worse. After the 1868 earthquake, a commission was created, but their report never saw the light of day. It must have had surprises. Our biggest surprise in a Hayward fault earthquake could be the loss of the Delta. It's not likely, but I don't think you can rule it out.

—William Lettis, of William Lettis & Associates

San Francisco, 1868. Damage at the northwest corner of Bush and Market streets from an earthquake on the Hayward Fault, more than fifteen miles away. (Unknown photographer, Courtesy National Information Service for Earthquake Engineering, University of California, Berkeley)

Los Angeles, about 1870. Commercial Street from the corner of Main Street. (University of Southern California Regional History Collection)

The Marina District in San Francisco, 1912. Shortly after this photograph was taken, the cove in the background at the right was filled with sand taken from the bottom of the bay to create land for the Panama–Pacific International Exposition. *(Unknown photographer, Courtesy National Information Service for Earthquake Engineering, University of California, Berkeley)*

The fire that resulted from the 1906 earthquake in San Francisco caused more damage than the ground shaking. *(Arnold Genthe, Courtesy National Information Service for Earthquake Engineering, University of California, Berkeley)*

Destruction on Mission Street after the 1906 earthquake. (*Doc Rogers, Courtesy National Information Service for Earthquake Engineering, University of California, Berkeley*)

Bird's-eye view of the ruins of San Francisco from Captive Airship, 600 feet above Folsom, between Fifth and Sixth streets, taken on May 16, 1906. (*Geo. R. Lawrence/The Library of Congress*)

Above: An apartment building in San Francisco's Marina District that was thrown off its foundation during the Loma Prieta earthquake on October 17, 1989. It had a magnitude of 7.09. *(Vince Maggiora/San Francisco Chronicle)*

Right: A collapsed section of the Cypress Viaduct of Interstate 880 in Oakland, after the Loma Prieta earthquake. Earthquake damage caused the double-decked freeway to "pancake." *(Tom Levy/San Francisco Chronicle)*

Left: A roadbed collapse on the eastern section of the San Francisco–Oakland Bay Bridge after the 1989 Loma Prieta earthquake. (C. E. Meyer/United States Geological Survey)

Below: Damaged and burned structures at Beach and Divisadero streets in San Francisco's Marina District, after the Loma Prieta earthquake. Fires were caused by broken gas mains. (C. E. Meyer/United States Geological Survey)

Above: Stranded vehicles sit atop a broken section of Interstate 5 near Sylmar, after the earthquake in southern California on January 17, 1994, which had a 6.69 magnitude. *(Brant Ward/San Francisco Chronicle)*

Left: A geyser from a water pipe that broke during the Northridge earthquake. *(Unknown photographer, Courtesy National Information Service for Earthquake Engineering, University of California, Berkeley)*

Separation of the freeway at the Interstate 5/Highway 14 interchange in the aftermath of the earthquake in Northridge on January 17, 1994. (Unknown photographer, Courtesy National Information Service for Earthquake Engineering, University of California, Berkeley)

A parking structure on California State University's Northridge campus, after the earthquake. (Unknown photographer, Courtesy National Information Service for Earthquake Engineering, University of California, Berkeley)

The Kaiser Permanente office building in Granada Hills, California, following the Northridge earthquake. The second floor has completely collapsed, and the bays at either end of the building have partially collapsed from the second to fifth floors. (Unknown photographer, Courtesy National Information Service for Earthquake Engineering, University of California, Berkeley)

A collapsed span of the Hanshin Expressway, following an earthquake with a 6.69 magnitude that occurred in Kobe, Japan, on January 17, 1995. *(Christopher R. Thewalt, Courtesy National Information Service for Earthquake Engineering, University of California, Berkeley)*

A tipped-over portion of the Hanshin Expressway. *(Christopher R. Thewalt, Courtesy National Information Service for Earthquake Engineering, University of California, Berkeley)*

We take the earthquake threat to the Delta very seriously. We just don't believe the mass levee failure hypothesis. We expect to be able to continue delivering water south from the Delta even if there's some levee failure. We had six islands flooded in the great storm in 1986 and we continued to deliver water.

—Les Harter, Chief of the Division of Engineering, California Department of Water Resources

MWD is so worried about the stability of the Delta in a major earthquake that we've done quite a bit of computer modeling to play with the consequences. The worst case would be mass levee failure during an intense drought episode like we had in 1976 and 1977. Based on our models, we might lose our northern California water supply for as long as three years.

—Paul Teigan, senior engineer with the Metropolitan Water District of Southern California

No water supply system in the U.S. is designed for wildland fire or the kind of conflagration we can anticipate following a major earthquake. After an earthquake, there are going to be many pipe breaks resulting in a loss of pressure, and there are going to be many fires. In dry conditions, especially in the East Bay, where homes spread all over the wildland-urban interface, one can only think "conflagration." In the worst, worst case, it's autumn and we have a strong Santa Ana type condition going. That's an unimaginable, scary scenario. That's your worst nightmare.

—Charles Scawthorn, Senior Vice President with EQE International, a prominent risk-management consulting firm

The basic lesson in life is: "You can't hide." The earthquake will impact you directly or indirectly. You will essentially be on your own for seventy-two hours. Only you, your family, and your neighbors can provide assistance. You should plan ahead to rescue people in damaged structures, to protect against fire, and to provide food and water for a minimum of seventy-two hours. You cannot *count* on *any* assistance or help from outside, although such help may, in fact, arrive, depending on the circumstances.

—*William Lettis, of William Lettis & Associates*

For disaster planning purposes, the five-foot average displacement along the Hayward fault will heavily damage all major tunnels and aqueducts that cross the fault zone. Similar damage will occur to the many treatment facilities and distribution water lines that cross the fault. The flow of water crossing the fault will be reduced to 10–30 percent for the first 24 hours. . . . Restoration of full service could take months.

—Earthquake Planning Scenario for a Magnitude 7.5 Earthquake on the Hayward Fault in the San Francisco Bay Area, *a 1987 report published by the California Division of Mines and Geology* (CDMG)

Due to loss of public utilities, reduced public confidence in structures in and near the surface rupture, and to access problems, all [eight] hospitals in or within one mile of the fault zone are closed and the patients transferred elsewhere.

—*The* 1987 CDMG Scenario

> Transmission towers and lines are principally subject to damage through secondary effects such as landslides and other ground failures. Conductor lines swinging together . . . could cause many burndowns. . . . While the loss of a few towers would not pose a formidable situation, damage could be widespread and significantly compounded by landslides during the wet winter season or by fire caused by fallen lines during the dry season.
>
> —*The* CDMG Scenario

> The Bay Area has the highest density of faults per square kilometer of any urban area in the world.
>
> —*William Lettis, of William Lettis & Associates*

The California Office of Emergency Services hasn't called the Hayward fault the most dangerous in the United States, but it does consider it "probably the most built-upon fault in the world." Other faults might compete for that distinction, but not the San Andreas. The southern segment of California's most famous fault makes a wide bend around the Los Angeles Basin, veering off into the Mojave Desert. The northern section heads offshore a few miles south of San Francisco and returns to land under the North Coast, where few people live. One could call it miraculous that a fault of such length (700 miles) in such a populous state manages to avoid most of its population. The San Andreas fault has proven that it can serve up a great earthquake—at least an 8.0—and tectonic upheaval of that magnitude, even ten miles from a built-up area, can cause a great deal of destruction. But a smaller fault can do a lot worse if it cuts through packed civilization like a highway stripe.

The Hayward fault's northern end is somewhere under San Pablo Bay, a couple of dozen miles north of downtown Oakland. It comes ashore at Point Pinole and runs as a discrete crustal rupture for sixty miles, going south-southeasterly before it feathers into complex dead-end fractures a few miles east of San Jose, beneath a mountain named Misery. Four and a half million people live on a strip of terrain thirty miles wide between its northern and southern ends, even though much of that landscape is covered by San Pablo and San Francisco Bays. These are sixty of the most populous, industrialized, infrastructure-dependent (eight great bridges, among other things), economically valuable, strategically important miles in the United States.

The fault goes along the eastern side of San Pablo, a big deteriorating industrial suburb. Then it goes through or alongside Richmond, a bigger deteriorating industrial suburb. Much of the housing stock dates from the Kaiser shipbuilding era—the late thirties to the sixties—although most commercial buildings are newer. The population is poor, mainly black and Latino, and few of these homes—flimsy single-story structures of wood or masonry—have been seismically reinforced. Richmond, whose population is about a hundred thousand, is one of the largest oil ports on the western coast of the United States, and has chemical industries producing huge quantities of toxic compounds.

(Every now and then, a leak at some plant forces thousands of people to evacuate.) Big refineries sit around the outskirts of the town, joined by pipes to several tank farms. A typical oil storage tank contains a million to five million gallons of crude or refined oil. The Hayward fault is a rifle shot away from a number of tanks, which are protected by dikes; if their contents spill and something sets them aflame, the conflagrations are not expected to spread (if it is not windy).

As it continues south, the Hayward fault crosses Interstate 80 and just misses San Pablo Dam, which holds back a reservoir in the East Bay hills large enough to wipe off a fair portion of San Pablo. The dam's durability during an earthquake was a great worry before it was recently re-engineered; now, most engineers believe it will survive. The fault runs over one set of East Bay hills, bisecting Contra Costa College, a popular community campus, before it settles in along the western piedmont, going through eastern sections of El Cerrito, Albany, Kensington, Berkeley, and the UC–Berkeley campus.

According to Raymond Seed, a well-known member of the Berkeley civil engineering faculty, the Berkeley campus and its southern California counterpart, UCLA, are two of the densest amalgamations of large, seismically vulnerable, unreinforced masonry buildings in the entire state. Their only rival may be San Francisco's Chinatown. Berkeley's Memorial Stadium, where seventy thousand spectators attend the Big Game between Cal and Stanford, is directly on the fault—not miles, yards, or inches from it. If, during an earthquake, seven feet of displacement should occur there, then one part of the football field will suddenly be one hundred two yards long. The stadium itself is apt to be a wreck. Fault creep has already contorted the structure by thirteen inches since it was built in 1924, requiring the installation of expansion joints.

The next strategic highway bisected by the fault is Route 24, ten lanes of east–west traffic going to and from the ballooning suburbs in the outlying East Bay: Concord, Lafayette, Walnut Creek, San Ramon, Danville, Blackhawk, Brentwood. The fault goes perpendicularly through the freeway. If the maximum credible surface displacement of ten feet should occur there, drivers in the eastbound and westbound inner lanes might find themselves aimed right at each other. Next to Malibu, the area where the fault crosses Route 24 may be the most disaster-prone piece of terrain in California. It is where twenty-five hundred structures burned during the 1991 Oakland Hills fire, and where several hundred homes a couple of miles north were incinerated in the Berkeley fire in 1923. The Concord Line of Bay Area Rapid Transit—BART—tunnels through the Oakland hills alongside Route 24's Caldecott Tunnel, and the fault surgically severs the western end of the BART tube. Immediately before, it rips within ten feet or so of Pacific Gas and Electric's Claremont substation Station K, which feeds electricity to much of the area, and swipes the eastern side of Lake Temescal, a small reservoir whose water could be of some value fighting fires. Then, as it continues south, it goes under the Warren Freeway for much of its ten-mile length. The cleavage from episodic faulting left such a perfectly straight rift near the base of the hills that California Department of Transportation (Caltrans) engineers couldn't resist laying a freeway in there; it saved them a great deal of excavation and expense—then.

The Warren Freeway, running astride the fault, travels east

of Piedmont, the East Bay's most elegant suburb. Piedmont is a showcase of mansions—Mediterranean, Tudor, Greek Revival, Federal Period, Italianate, Bauhaus Cubist—and rambling Arts and Crafts cottages designed by Julia Morgan, who was the Bay Area's most famous residential architect, and her disciples. Piedmont, as the name suggests, sits low along the East Bay hills; its well-built homes may stand up to strong shaking, and the worst landslide danger is higher up; however, with its tall pines and eucalyptus trees and many wooden homes, Piedmont is susceptible to fire.

As the fault veers by Piedmont—which, unlike most neighborhoods, is heavily insured—downtown Oakland is four to five miles away.

The public infrastructure bisected by the Hayward fault, or sitting nearby, is awe-inspiring. The Richmond–San Rafael Bridge and the Bay Bridge are about five miles away. The eastern sections of the San Mateo and Dumbarton bridges are closer. The main highways around the southern end of San Francisco Bay and the northern end of San Pablo Bay cross the fault or come near it, and long sections are on fill. Interstate 580, one of the main East Bay freeways, is crossed three times by the fault. No fault in the world is apt to destroy so many vital transportation corridors. In fact, there is hardly a Bay Area lifeline that the Hayward fault *doesn't* threaten. The main 230-kilovolt lines feeding electricity into the entire region cross it or come near. San Francisco's Hetch Hetchy Aqueduct is bisected by the fault, and the terminus of the East Bay's Mokelumne Aqueduct is a stone's throw away. From East Bay Municipal Utility District's (EBMUD's) Claremont terminus, big mains, smaller connector mains, and hundreds of feeder mains branch out across and along the fault (over 230 separate pipes in

EBMUD's system cross the Hayward fault). The same applies to gas mains, to sewer lines. Communications cable is laid all over the fault zone, putting the region's telephone and computer systems at risk of mass failure. The Oakland Airport is three miles away. San Francisco International Airport is twelve miles away. Hayward Municipal Airport is practically on the fault. Alameda Naval Air Station—decommissioned, but serviceable in any other emergency—is four miles distant, on abysmally poor soils. There are eight hospitals within a mile or so of the fault trace, where violent ground motion—well over 1.0g in some locations (the 1994 UBC specifies 0.4g in this area)—should occur. San Leandro's Lakewood Hospital, like Berkeley's Memorial Stadium, straddles the fault; it's so vulnerable that a few of its worst buildings have already been removed. So many homes are virtually atop the fault, and so many of their owners and renters are new to California that many people who sleep on it every night have never heard of it. (After half a dozen people who ought to know told me this, I finally believed it.) South of Hayward, a city dating back to the 1850s, you have Fremont, a sprawling suburb that is, even by California standards, brand-new. Fremont is a major headquarters of warehousing and light manufacturing, absorbing a lot of spillover from Silicon Valley. The fault cuts through the middle of it. The original Silicon Valley, just across the bay, has planted a lot of buildings, and tens of thousands of workers, in soft soils. Housing prices in the South Bay are so humiliating—in 1999, half a million dollars bought you a small postwar three-bedroom, perhaps two thousand square feet—that some technology companies have gone elsewhere. There is a Silicon Valley plasmid in Emeryville, north of Oakland—where Chiron, a leading biotechnology firm, has its headquarters. High-tech firms have come to bleak, crime-ridden Richmond; across the bay, they've sprouted in San Rafael. Industrial Light and Magic, the wildly successful special-effects studio founded by the billionaire filmmaker George Lucas, has grown up there, on a stretch of bay margin once favored by egrets, herons, and ducks. In 1999, ILM had a force of eleven hundred cybergeeks filling eighteen buildings. They sit at incredibly expensive imaging computers and conjure artificial anything: hurricanes, blizzards, explosions, vehicle-eating dinosaurs, tidal waves that topple skyscrapers. Earthquakes. Few of the people who work there may know that the complex sits partly atop man-made terra firma that, in a real earthquake, will perform its own special effect: it will liquefy.

I had lived in the Bay Area for nine years before I knew anything of the Hayward fault. It was 1987, and I had just published my first book, *Cadillac Desert*, a history of water and the American West. Although various critics had praised the book as "exhaustive" (a compliment that can hurt sales), one wrote me a letter elaborating on what, in his view, was a gargantuan omission.

His name was Michael Finch, and he was a geologist then employed by the California Department of Water Resources (DWR). Finch's area of expertise was the Delta, the huge reclaimed marshland at the confluence of the Sacramento and San Joaquin Rivers east of the Bay Area. The gist of his criticism was that I had said nothing about the Delta's vulnerability to strong ground motion during an earthquake on a nearby fault.

Finch seemed obsessed with such an eventuality—so much so that he had written a three-hundred-page manuscript on the subject, which he later mailed to me. His hypothesis went like this: If, during a strong nearby earthquake (he didn't have just the Hayward fault in mind), a number of Delta levees were to slump and collapse through liquefaction of underlying soils, the region might undergo mass inundation—an event that could paralyze the *southern* half of California, too. No fault in the Western Hemisphere was apt to cause such far-flung effects.

For more than a century, Finch wrote, Delta farmers had been working the most oxidable, and ephemeral, of soil types: peat. As a result, an area thirty times the size of Manhattan had subsided ten to twenty feet below sea level. What you had there now was a vast empty reservoir, a man-made hole in the California landscape. With levee protection lost, the below-sea-level Delta would become, in effect, a vacuum, which nature abhors. Water would pour in there as it would down a manhole. A lot of it would be saltwater sucked in from the bay. If a strong tide was pushing in when the levees failed, things would be that much worse. If it was summer or fall and freshwater outflow through the Delta was meager, it would be worse still. Half of the water supply of greater Los Angeles, San Diego, and the San Joaquin Valley goes through the Delta. Two-thirds of greater San Jose's—which is to say, much of Silicon Valley's—water comes from the Delta. Within hours or days, all that water would be unusable and undrinkable until the incursive seawater was pumped back out.

But how? Repairing the damaged levees would take months. It could take *years,* if it was even possible. Wind-driven waves on this new arm of the sea—huge brackish lakes separated by sodden dikes—would erode intact levee sections as fast as the broken ones could be repaired. Delta levees are more vulnerable to erosion on their inboard sides, which haven't been riprapped or otherwise reinforced.

How would you avoid a calamity like this? Finch's assignment with DWR had been to assess Delta soils, the ground on which the levees sit. In many places, the underlying strata are a mixture of semi-saturated silt and sand, highly susceptible to liquefaction. There really is no way to make the levees failure-proof, he argued. You could make them wider, you could make them higher, but during a strong shake they still might fail; re-engineering might make them fail more abjectly, due to the added weight. The only way to keep fresh water flowing reliably south was to bypass the whole region—to divert the water into a canal that circumnavigated the whole precarious artifice of crumbly levees atop unstable ground. In other words, build the Peripheral Canal—the same huge connector that northern Californians had for decades ferociously, and successfully, opposed.

The loss of the main water supply for twenty million Californians—not to mention irrigation water for four million acres of the world's richest irrigated cropland—seemed like a science fiction scenario. But, according to Finch, it was possible if the quake occurred on one of the Bay Area faults, and *probable* if it occurred on a fault underlying the Delta. Geologists were still uncertain how many faults ran beneath the region, so it was hard to guess what kind of risk they posed. But several levee sections had shown signs of fracture and slumping after the 1983 Coalinga earthquake, which was only a 6.5—on a previously unknown fault—and the epicenter was a *hundred forty miles south* of the Delta. What would happen during a quake ten times as powerful on, say, the Hayward fault, thirty-three miles from the Delta's western edge, or on some blind thrust fault right below the Delta?

I remember being stunned by Finch's scenario. Then I forgot all about it. Daily life does this to you in California. Besides the quotidian distractions you encounter anywhere, there has been, since I moved here in 1978, such serial disaster—famously destructive floods in that and the following two years; bigger floods in 1982 and 1983 and then 1986; the worst drought on record (accompanied by great wildfires and a billion-dollar citrus freeze) on the heels of the 1986 floods; then epic mudslides when torrential rains returned in 1992; then southern California's wildfire siege in 1993; huge floods again in 1997 and 1998—that speculative calamities don't count for much. But when earthquakes cast their spell over me, Finch's Delta hypothesis began camping out in my head. He had finally found a publisher for his manuscript, but sales were inconsequential. A lot of Californians don't know where, or what, the Delta is. After we spoke a couple more times, Finch abruptly left DWR; I tried to find out where he had gone, but no one seemed to know. Not long afterward, I arranged a meeting with some DWR officials to talk about the Delta and, incidentally, try to learn of Finch's whereabouts. Finch's former division chief insisted, adamantly, that his departure had nothing to do with his "Delta Doomsday" hypothesis—which, he happened to believe (vigorous head-nodding from everyone else), was a Gothic exaggeration, an excursion into fantasy.

When I shared this experience with a couple of earthquake experts who had told me Finch's prophecy might be close to the mark, one was particularly mystified. "These guys have just deliberately stuck their head up their ass," he said. (He has some experience in the arena of California water politics.) "Finch was bringing up a nightmare scenario which, if halfway credible, they should have been moving mountains to avert. Actually, to be fair to those guys, they've been trying to build the Peripheral Canal for the last thirty years. But northern California, especially the Bay Area environmentalists and the Delta farming interests, have always managed to kick their butts. To most northern Californians, even those who can't stand environmentalists, the canal is anathema in its purest form. It's a theological issue. The Peripheral Canal lobby *raised* the Delta Doomsday scare in 1982, in the propaganda they put out to get Californians to vote for the initiative that would have funded the thing. If I'm not mistaken, DWR people who call Finch a scare-monger today sounded a lot like Finch back then. But they lost that referendum vote by . . . what? . . . a 60–40 margin statewide. In the Bay Area, people voted nine to one against the canal. I get the sense that DWR sort of gave up after they got slaughtered in '82. If you're Dave Kennedy [then the director of DWR], and you spent your life trying to build the Peripheral Canal and even *southern* Californians oppose you in pretty fair numbers, maybe you just tell your fellow citizens to go screw themselves. You ask God to let the Delta collapse on someone else's watch. But, meanwhile, an alarmist like Finch is a nuisance. Bureaucracy hates a nuisance."

Suppose the Delta Doomsday hypothesis is credible? For months I had been wondering which big earthquake, among those considered likeliest to occur, could extract the greatest revenge for all the liberties we Californians have taken with the natural order of the state. If the Hayward fault were to produce a quake that wrecks the Delta, the Bay Area would be paralyzed, southern California water-starved. Any fantasized calamity, especially one so far-reaching, is apt to get a lot of things wrong. Several Hayward earthquake "scenarios" have been published—the most current by the Earthquake Engineering Research Institute in 1996, based in Oakland—and as valuable as they are for their scientific credibility, they're all vague and heavily qualified. There will be "major" power outages of "unknown" duration, there will be "substantial" loss of life, there will be "serious" structural damage in "poorly engineered" buildings—what does any of this tell you? The horror and chaos, the magisterial destruction from such an immense natural force—none of it is palpable. So I have decided, quite recklessly, to take the leap into fictional reality, leaning on expert opinion but helping myself to literary liberty.

What I am about to conjure is not what scientists call the "hypothetical worst case." My earthquake is a 7.2, high up in the expected Richter-scale range of intensity; most seismologists say a 7.0 is likely, a 7.2 merely possible. It strikes in 2005, which is at the early side of recurrence predictions. Planned re-engineering projects of supreme importance—especially a new eastern section of the Bay Bridge—are not likely to be finished by then.

I am now writing on October 20, 1999; tomorrow is the twenty-first. On the same day in 1991, ambient humidity in the Oakland hills was 8 percent, it hadn't rained since spring, and, above the Caldecott Tunnel, a dying campfire set by a homeless man was rejuvenated by sixty-mile-per-hour Santa Ana winds. Those embers became the inferno that obliterated three thousand East Bay homes. At least a hundred twenty *substantial* fires are expected—just in the East

Bay—after a Hayward fault earthquake. And much of the region will have lost its water supply. Conversely, I could stage the event during a winter like 1997–98, when the same steep terrain—sprouting many new homes, but sparsely revegetated and primed for landslides—got a sixty-inch soaking in three months.

What follows is pessimism that could be worse.

It is February 28, 2005, an unseasonably warm and sunny late winter day toward the end of a moderately wet winter season. The first shock was registered at six seconds past 2:38 in the afternoon.

All over the Bay Area, seismograph needles, careening over phlegmatic cylinders, etch a dizzying graphic record of needlepoint peaks and abject troughs. On the Berkeley campus, in Davis Hall, the seismographs are already being crushed as the building loses a wall, then collapses toward its middle with students and faculty inside. Computerized sensors point to the middle or northern end of San Pablo Bay as the epicenter.

The noise emanates from the basement of time, a Cyclopean orchestra tuning instruments. It rolls in on a basso profundo roar. (During Loma Prieta, it began with a sonorous, bottom-octave moan, steadily blending with the sounds of things straining, fracturing, cracking, settling, popping, and shattering; but what I remember most vividly is the grinding, the unearthly noise of great surfaces and structures grating together.) You are not prepared for the intensity of the shaking. You aren't prepared for how it strengthens. It takes two or three seconds to sink in: an earthquake! *An earthquake!* AN EARTHQUAKE! By the time it does sink in, ground motion has become upheaval—you can hardly stay on your feet. An epiphany dawns: *This is It! The One!* Surprise yields suddenly to shock, and to atavistic dread. "We'll be scared to death in a quake of this magnitude," Rich Eisner with the Office of Emergency Services (OES) had told me several years earlier. "We won't just be saying, 'My God, it's an earthquake!' It will be the most terrifying thing most people have ever experienced."

People rush out of buildings like bats from a cave. They cluster under trees or run for open space, where things aren't apt to land on their heads. In buildings too big or too tall, where fleeing is pointless, you can only ride it out.

The epicenter is eight miles above the hypocenter, the juncture in shallow planetary crust where the Pacific and North American Plates began to separate. The Pacific Plate has lurched northward; the North American Plate has more or less stayed put. A colossal bolt of kinetic energy races from the hypocenter, like the shock wave from an atomic bomb. (The energy released by this earthquake equals that of thousands of Hiroshima bombs.) The initial energy wave, the P or primary wave, broadcasts from the hypocenter at roughly fifteen thousand miles an hour. The trailing wave, the secondary or S wave, follows at up to half that speed. As they peel off the hypocenter, the two source waves segment into a series—a set, like tremendous ocean swells—that stretch out and lengthen the duration of strong shaking. Each wave set causes distinctively different kinds of ground motion. The P wave pushes terrain forward, releases it between pulses, pushes again—the motion is back and forth. The S wave slams from side to side. No matter—the ground seems to lurch in all directions at once. Where the propulsive energy of the seismic waves rips through softer soils, it slows. The ground motion is amplified and lengthened more.

Around the margin of San Francisco Bay, you can see the waves coming—the earth's surface bulges and flops like a tarp. The shaking is so violent that it wipes people off the ground and bounces cars in the air.

It doesn't stop. Fourteen seconds. Nineteen seconds. Amid the preternatural, blended din, thousands of motion-triggered car alarms go off. In San Francisco, skyscrapers sway frighteningly. A few windows pop out of tall buildings and bomb the streets below. In Golden Gate Park, some brittle Monterey pines shear their trunks. On the waterfront, a pair of great old docks slump into the bay as their pilings are kicked out. Bricks explode from Chinatown buildings, and their walls begin to fall.

The roadway of the Golden Gate Bridge is rising and falling seven feet as vast pulses of seismic energy sear through Fort Point and the Marin Headlands. Because the directivity of the energy pulses is perpendicular to the bridge, the suspended deck, which has a great deal of transverse give, is also rocking sideways. The deck's expansion joints, interlocking sets of steel grates directly under each tower, were installed to let the three deck sections—the one between the two towers, and the other two at either end of the bridge—flex without crushing one another. With an unearthly creaking and moaning, the joints open and close like barber's shears.

Because the Golden Gate Bridge has a morning inbound and an afternoon outbound rush hour—but has only a single six-lane deck—traffic moving in either direction may use two, three, or four lanes, depending on the time of day. A movable, but relatively crash-proof, north–south traffic divider has been talked about for years and various prototypes have been designed; but in 2005, years and several head-on crashes after a divider was first proposed, it hasn't been installed. Vehicles in the north- and southbound fast lanes, generally going fifty to sixty (the speed limit, not so strictly enforced, is forty-five), are separated by nothing more than slim rubber pylons a couple of feet high. The lanes are notoriously narrow, and cars headed in opposite directions are barely an arm's length apart. Unnerved by a bridge that suddenly sways like a swing, several drivers stray momentarily into neighboring lanes. A light truck in the outer southbound lane sideswipes a car; the panicked driver veers sharply away, begins to spin around, overcompensates, and whirls off into oncoming traffic. The pickup crushes the front of a Mazda Millennia, flips all the way over it, and continues skidding upside-down into oncoming vehicles—a grisly grind-up that sprays sparks, metal pieces, and a portion of a human arm across the bridge deck. Two northbound drivers veer toward each other, unconsciously, to avoid a rolling tire and hood that has skidded in front of them; the impact sends them spinning out of control. Even though traffic is braking furiously, cars continue to smash. There is wreckage across all six lanes under the south tower—ten cars are badly mangled. Even after the shaking has stopped and traffic on the deck has stopped dead, the bridge continues to sway rhythmically. People abandon their cars and sprint for the plazas at either end.

Along Interstate 80 at the Berkeley waterfront, a stretch laid down over landfill about five miles from the fault, accelerometers record peak surges of 0.45g. The ground motion is relentlessly violent. Seconds after the shaking starts, the pavement begins to tear apart as the subsoil liquefies, slumps, and spreads laterally toward the bay. Great cracks open in the asphalt. Sand boils eight feet high erupt from the savaged roadway, and black goo geysers into the windshields of speeding cars. The two right-side southbound lanes, slumping quickly, drop two feet below the three left lanes. The breakdown lane becomes a gulch—it simply disappears.

Drivers go bug-eyed as a ten-lane interstate under their wheels shifts westward, spews stinky black mud, and rips apart like a piece of cloth. Those not too stupefied to react fight to control their bouncing, swerving cars. Vehicles slam into holes and break axles. Tires lock into pavement fissures. A rear axle separates from a Jaguar and as the undercarriage slides across the highway, expelling showers of sparks, the free axle crashes into an adjacent light truck and punctures its gas tank. A spark ignites the escaping fuel: a dull explosion, then streams of fire spread over five lanes of traffic. Panicked drivers try veering around it; they sideswipe cars next to them, spin out of control, come to a rest facing oncoming traffic. There are two mass collisions on the heaving, sinking, dissipating waterfront reach of I-80—they occur almost simultaneously between the Bay Bridge approach and the junction with Interstate 580, six miles north. That whole stretch of heavily traveled freeway will be undrivable for weeks. For now, it's strewn with wrecked cars, which are filled with injured and dead. As violent earth movement and liquefaction macerate other stretches of bayfront freeway—U.S. 101 near the San Francisco Airport and Candlestick Park, Route 84 east of the Dumbarton Bridge, Route 37 hugging the northern shore of San Pablo Bay, Interstate 880 at Alameda—the region's main transportation corridors are knocked out as if by carpet bombing. Some drivers, desperate to drive on, rattle slowly across the bombed-out stretches; others are too frightened and simply sit in their lanes, blocking the stream of traffic coming up behind them. Many drivers pull over to the breakdown lanes, a deadly obstruction to police cruisers and ambulances that will soon try to thread their way to the accidents. There are roughly six million vehicles in the Bay Area, and a million or so were in transit when the earthquake struck. Traffic jams, many as much as seven miles long, won't clear for two days.

Downtown in Oakland, City Hall's ornate neoclassical tower, twenty stories tall, rolls and lists like a buoy at sea. It is riding a hundred-million-dollar retrofit, a base-isolation project completed in 1995 that hydraulically hoisted the whole structure and set it back down atop massive rubber-and-steel mounts. The building was doomed before the retrofit, but now it performs exceptionally; it rides the pitching terrain like a ship. The tower's clock begins to rip from its mounting, but it doesn't fall. City Hall is the exception; all around is horror and chaos.

San Francisco's downtown area was cleansed of most older, fragile masonry buildings during a building frenzy that lasted from the late sixties to the later eighties. Some treasured architectural landmarks were lost and San Francisco's downtown core ended up looking like that of any other big American city. But a lot of unreinforced masonry buildings were replaced by skyscrapers mounted on stronger foundations, skyscrapers whose endoskeletons and exoskeletons are far more flexible and able to shed seismic energy. In Oakland, a downtown renaissance gathered momentum in the eighties, then faltered. Peer down any number of core-district streets, pretend that the occasional new twenty-story office tower isn't there, and you have a downtown preserved from the twenties. Old brick-fronted masonry buildings stand by the score. Many are cheap hotels, rooming houses, and office buildings filled with welfare families or members of a struggling, aspiring new immigrant class that has transformed wasted blocks into microcosms of Seoul, Manila, Taipei, and Shanghai. The fringes of downtown's core are rife with such building stock—the presidential streets between Twelfth and Eighteenth; Martin Luther King Way; Telegraph and Broadway as they bore into downtown from Berkeley. After Loma Prieta, which leveled similar buildings in Santa Cruz, Oakland demanded that owners of buildings susceptible to complete failure perform seismic retrofits. The owners balked, saying their low rents and valuation didn't justify half-million-dollar re-engineering projects. Some of them complied, but sixteen years later many still hadn't; city officials sent them warning after warning, but took the landlords seriously when they threatened to walk away from their properties. Oakland couldn't afford to take them over, and didn't want thousands of new homeless on the streets.

Just beyond the downtown core, around the northern fringes of Lake Merritt, are newer but equally flimsy structures, solid blocks of four-story apartment buildings from the Moderne era now occupied, in large part, by retirees.

All these sections take a tremendous hit. As City Hall and the new high-rises sway breathtakingly, buildings up and down the street implode. Even those that were re-engineered after Loma Prieta aren't performing well. Cornices scab off, ornamentation shakes loose. One hotel near

Martin Luther King and Fourteenth emits a ghastly creak as a middle floor sags, bowing the structure like a derelict old barn until all five floors cave in.

Simultaneously, a score of older downtown buildings are cracking apart. With a horrible shattering and shower of glass, two newer glass-walled mid-rises lose their first floor: their steel frames crumple between the first and second story and the upper floors fall onto the first floor—exactly what happened in Kobe. An incredible cascade of rubble flows into the streets, carrying building occupants with it, burying others who have just run outside. Shrieking crowds flee toward City Hall's little park, the new Federal Center plaza—any place where they're less likely to be buried. Some people climb fences surrounding vacant lots and hurl themselves toward safety. The landsliding buildings fall not just over humans but dozens of cars—some that were parked, others that were driving along when they got crushed.

In a single minute, downtown Oakland has become Sarajevo, or Chechnya. There is no fire, yet. Not far from Lake Merritt, a steel-and-glass office building, nearly twenty stories tall, is tilting sickeningly. It has a huge soft space at ground level, a big bank branch office with a twenty-foot ceiling and too little columnar lateral support; the shaking must have weakened some structural joints in the steel frame enough to cause the upper eighteen stories to collapse—at least partially—onto the soft story below. Perhaps it was a combination of soft story on soft soil. It is surreal—a tall modern building, still filled with people, leaning at a twenty-degree angle. People are trying to get out, but the doors are squeezed shut—the doorframes contracted under the force of so much dislocated weight. Finally, a security guard heaves a typewriter through a glass pane, and people begin to scramble outside.

The body count in Oakland's downtown core will be horrific—the worst anywhere. In Berkeley, which also has clusters of unreinforced masonry buildings, it will be awful too—probably even worse than on the Bay Bridge.

During Loma Prieta, a short fifty-foot section of the bridge's upper deck (the entire length of the structure is double-decked) snapped off its anchoring atop a wide pier known as E-9 and dropped like a trapdoor onto the lower deck. The force of the impact nearly unhinged the lower deck, too, and the whole section came precariously close to slipping off its pier and plunging a hundred thirty feet into the bay. But it held, with two mangled cars dangling over the gap. A passenger in one of the cars was killed when the vehicle dove from the upper deck onto the lower.

Among the various consulting engineers hired by California's Department of Transportation to investigate the E-9 deck failure, the common reaction was surprise—surprise that more of the bridge didn't fail.

The Bay Bridge, which, like the Golden Gate Bridge, was erected in the mid-thirties, is less celebrated than its regal, magenta-painted cousin a few miles to the west. But contemporary engineers might applaud its forgotten builders more than they would Joseph Strauss, the egomaniacal chief engineer who let no one forget that he bridged the Golden Gate. For one thing, the Bay Bridge is nearly five times as long as the Golden Gate Bridge; in its day, it was the longest true bridge in the world, by far. It also carries more traffic—

about a quarter of a million vehicles each weekday. There are five bridges crossing San Francisco and San Pablo Bays, and the Bay Bridge handles as much traffic as the other four. What's most peculiar, though, and most ingenious, about the Bay Bridge is that it is several distinct bridges formed into one: two great, joined suspension spans and a complex series of cantilever and truss spans that stretch from Yerba Buena Island to the Oakland shore.

The western half, the suspension span, which vaults majestically off San Francisco's bayfront skyline, is nearly two miles long—much too long, at least when it was built, for a single span like the mile-long Golden Gate. But the ship channel that the bridge crosses, between Yerba Buena Island and San Francisco, was already being used by aircraft carriers when the bridge was planned, and those vessels were certain to get bigger and bigger—and higher. High clearance was essential, and that argued powerfully for a suspension bridge. The only option was to build two of them. But two 4,500-foot suspension bridges required a second island, between Yerba Buena and San Francisco, to anchor an end of each one. The engineers had to create one. From many miles up or down the bay, you can see this outsize apparition—a concrete block with the mass of a giant dam, rising hundreds of feet from bedrock below the bay. It was, in fact, and may still be the greatest mass of concrete anywhere that isn't a dam.

The other half of the bridge, the eastern section—which stretches from Yerba Buena Island to the Oakland shoreline—looks less impressive, but it's an equally inventive piece of work. The first section is a cantilever truss bridge, with a high superstructure overhead. Then comes a complex segue of truss sections—five deep 50-foot-long simple-span trusses followed by fourteen shallow 289-foot-long simple-span trusses—as the bridge approaches the Oakland shore.

The last seven thousand feet before Oakland landfall was always the bridge's weak link. The muds are so deep and the depth to bedrock on that side of the bay is so great that anchoring the supports to bedrock wasn't feasible. The designers had to come up with another idea. One engineer whom I interviewed told me he laughed in disbelief when he learned, in graduate school during a seminar on bridge design, what it was. The above-water, concrete-and-steel piers that hold up both decks of the eastern stretch of the Bay Bridge are themselves fixed in place by sixty-foot, coopered-together bundles of logs. The eastern truss section, in other words, sits on nothing more than pounded-down pilings; it looks like a bridge, but it's really more like an inclined dock carrying cars. The pilings were sunk well below the bay bottom, down into dense, compacted sedimentary soil that could secure them firmly enough to handle the weight of a bridge deck, plus hundreds of vehicles.

But not violent shaking caused by a nearby earthquake.

At the bridge's eastern end, six 289-foot-long pieces of the truss section were mounted on half a dozen piers designated E-17 to E-23. Atop one pier—say, E-17—the end of a deck section was anchored by big bolts. The opposite end, resting on the next pier—E-18—was mounted on a series of rollers designed to accommodate steel's tendency to stretch or contract as temperatures change. The rollers would also let the bridge sections pass seismic energy by moving, or flexing; a too-rigid bridge would snap into pieces.

The pier tops were designed to accommodate seven inches of longitudinal flex. It isn't nearly enough.

After Loma Prieta, engineers inspecting the truss span for damage discovered that all of the bolts securing some sections to piers had sheared in half. That left segments of deck almost as long as football fields free to slide back and forth like boards resting on a pair of logs. Gouge tracks offered evidence that the end of one truss section came within an inch of falling into the bay. The bridge was out of service for a month while the fifty-foot E-9 deck was repaired; meanwhile, the vulnerable truss sections were rebolted to their supports with stronger hardware.

For hundreds of thousands of people, those thirty days were a commuting purgatory worse than the closure of the Santa Monica Freeway after it dropped an overpass during the Northridge quake. However slowly or chaotically, you *can* route traffic around a closed stretch of freeway, as long as you have parallel streets. You can't take vehicles off a five-mile bridge and dump them in a bay. When the bridge was down, BART trains operated at maximum capacity for the first time, and ferry service between Oakland and San Francisco, which was discontinued a few years after the Bay Bridge was finished, was hastily restored. But they couldn't handle all the displaced commuters—not to mention trucks—so tens of thousands of vehicles crossed on the other bridges, where they caused merciless jams.

Even after the repairs, the bridge's seismic reliability was only marginally improved. The new bolts were twice as thick as the ones that snapped, but still weren't expected to handle the energy release of a 7-plus quake on the Hayward or San Andreas fault. A magnitude 7.3 Hayward quake, according to computer models, would create seismic demands on these bolts *ten* times greater—and five times *longer*—than the moderate loading they got from Loma

Prieta. As each 6,000-ton truss section (plus vehicles) rocked back and forth, the massive weight would put intolerable pressure on the bolts. The same models suggested that only gigantic bolts—as thick as a thigh—could withstand superpressures like these. But by *not* shearing such bolts, and thus relieving stress on the bridge, seismic energy would be hurled back at the piers. In that case, slabs of concrete could shear off, warping the steel understory and causing the bridge to fail in an entirely different way.

The California Department of Transportation, which is responsible for repairs and maintenance of the bridge, seemed dumbfounded by this dilemma—it had limited expertise in complex steel bridges—so it recruited a fair-sized army of consulting firms to come up with a fix. One of them, a small but highly regarded group called G & E Engineering Systems, completed a report three and a half inches thick. G & E's models put the odds of a truss-section collapse at 95 percent during a very strong Hayward fault quake. Even in a moderate earthquake—just a 6.0 on the nearby Hayward fault—the truss section stood a fifty-fifty chance of failure of one span. "The more we experience strong earthquakes," G & E's affable president, John Eidinger, told me, "the more we appreciate the structural havoc they can cause. The guys who designed the bridge in the thirties did beautiful work, but they didn't have the benefit of Northridge, or Kobe, or Mexico in '85. They weren't bad engineers. They were innocents."

Eidinger's firm recommended a quick and cheap (which is to say, $60 million) retrofit of the eastern two thousand feet, between piers E-17 and E-23. If the bolts sheared again—and they would—strong shaking from a 7-plus Hayward quake could slide a couple of these deck plates three or four feet. They wouldn't roll off their piers—they'd fly off, and so would several dozen cars. G & E's simple answer was to build out the tops of the piers, widening them significantly, so that even a truss section shimmying forty inches back and forth wouldn't go off the edge. Other parts of the bridge would be damaged, but the truss section was going down for sure and should be fixed posthaste.

Caltrans, however, held out for a vastly more expensive (its initial estimate was $1,200,000,000) *completely new* bridge replacing both the truss and cantilever sections from Yerba Buena Island to Oakland. It was to be parallel to, and a few hundred feet north of, the old bridge, curving back into the Yerba Buena tunnel at the western end.

Brian Maroney, the youthful Caltrans engineer in charge of the project, was a product of Peoria, Illinois—a city where I also grew up, as an adolescent—and, even after living in California for some time, was still acclimated to midwestern like-mindedness and conformity, which midwesterners call community. He was not prepared for California's myriad "communities," which battle one another with fixed bayonets over issues that midwesterners might find absurd. No one from Peoria, Illinois, could foresee that, even in the case of an emergency bridge replacement project, dozens of organized interest groups would have militantly strong opinions about how it should look, where it should go, what sorts of traffic it should carry, and who should pay for it. Maroney, in the words of one individual well acquainted with California's mega-construction projects, remains "a breath of fresh air in state government."

Did anyone *not* have a serious concern? The environmental community was *seriously concerned*, among many other things, about toxins coming out of disturbed bay muds during construction, poisoning fish and wildlife. Elements of the mass transit community were *seriously concerned* that no light-rail line was part of any plan; they wanted one whether the bridge was rebuilt or brand-new. (There had been a trolley line on the Bay Bridge until the 1950s, when it was removed to make room for more cars; but it was soon replaced by a vastly more efficient mass transit system—BART—that crossed the bay in a tunnel right below the bridge. When BART ridership doubled after the bridge was shut down following Loma Prieta, it was ultimate proof that the region's commuter rail system was not yet operating at full capacity. So why restore an antiquated and expensive trolley line up the middle of the bridge—a line that was not only inferior to BART but would compete with it?)

Hard-core members of the bicycling community, who staged traffic-paralyzing "ride-ins" on San Francisco streets, were *seriously concerned* about the absence of a bike lane in every proposed design. (There is one on the Golden Gate Bridge, which few riders use. Also, if you built one on the new eastern section, you'd have to add one to the suspension span beyond Yerba Buena Island, and there was no room—unless you took out a couple of lanes in each direction, which is what some people wanted in the first place.) Bay Bridge commuters, a formidable lobby, at first resisted higher bridge tolls to help finance the retrofit; they demanded money from coffers filled by the state gasoline tax. Southern California commuters, a more formidable lobby, would have none of that.

A bewildering, boundless emotional energy was invested in the debate. As it dragged on and on, reporters seemed to grow bone-tired of the whole issue, but anyone who attended a public meeting knew that thousands of Bay Area citizens—*seriously concerned* citizens—were ready to fight to the end. By early 1999, finally, as the earthquake clock kept ticking, a design—a streamlined single-tower modified suspension span—was finally selected by a panel of officials after thirty-four public hearings had been held, over the course of two or three years, and more than $40 million had been spent on design engineering and public "outreach." In 1998, a one-dollar toll surcharge did go into effect, without incident, on all the trans-bay bridges except the Golden Gate, so the cost of the new span (now estimated at $1,500,000,000, a figure that was sure to rise) and of retrofitting existing bridges fell on those who use them most. Everything was settled.

Except that, in Bay Area politics, nothing ever is. In that same year, 1999, with the design, features, and route already selected by an impressive panel of experts, the Mayors Brown decided to take charge. Willie Brown, the former assembly speaker who had become mayor of San Francisco, was keen on developing Treasure Island—the big ovate chunk of landfill dumped next to Yerba Buena Island for the 1939 Exposition. It later became a military base, which then closed, and San Francisco was busy negotiating its acquisition from the navy. But the new eastern span, as designed and routed, would graze Treasure Island's southern edge, where some of Brown's business friends hoped to build a bed-and-breakfast and brew pub, among other things. The bridge would create *shadows* there. So the mayor suddenly insisted that the new bridge be built *south* of the old one, or San Francisco—in the mayor's mind, San Francisco and he were the same—would oppose the deal. The fact that a new routing could add much delay and at least $175 million to the bridge's cost, that it could stymie the Port of Oakland's expansion plans, that it would rip through a proposed new park at the waterfront, that it would conflict with huge sanitary sewer outfalls under the bay—none of this seemed to faze Mayor Brown. Then the Department of the Navy began to weigh in, insisting that it hadn't received *proper notification* about the thirty-odd public hearings Caltrans had convened. To demonstrate its pique, the navy, in early 1999, refused to allow Caltrans on Treasure Island to conduct some test drilling—a "naval blockade," in the words of Alameda County supervisor Mary King, that added $10 million to the cost of the new bridge every month. The navy also threatened that it wouldn't transfer Treasure Island to San Francisco until its conditions—which still weren't entirely clear—were met. (That would be a huge problem for Caltrans, a state agency that can exercise eminent domain over city property, but not federal property.)

Meanwhile, over in Oakland, the other Mayor Brown—Jerry Brown, the former two-term California governor and son of earlier governor Pat Brown—had a different objection to the new design. He didn't like its looks. It wasn't as *pretty* as the suspension span on San Francisco's bay. To correct that defect, the other Mayor Brown demanded an "international competition" among bridge designers to come up with a "world-class" structure. (Something much like that had already been held.) Like Willie Brown, Jerry Brown also asked the new governor, Gray Davis—who had been Brown's chief of staff when *he* was governor—to reconsider a new span south of the existing one. The governor, taken aback by the mayors' seeming indifference to the seismic threat to the bridge, demanded now and then that the Bay Area "get its act together," then disappeared on other business.

It went on like this for a couple more years, until everything was finally resolved, more or less. If all the politics, all the finances, all the permits, all the environmental review went correctly, the new bridge would be open in 2006 or 2007. Sometime in 2003, supports for the new bridge began to rise out of the water. When the earthquake struck, none of the deck was yet in place.

The ground acceleration recorded at Yerba Buena Island, near the bridge, is 0.7g—four times as powerful as in Loma Prieta. The strong shaking lasts five times longer, giving anything unstable a greater opportunity to collapse. A party of fishermen in a boat a few hundred yards off the Oakland shore said they could see the truss deck shift westward; they heard the sharp explosion of snapped bolts. In G & E's report, the E-17 pier, the seventh out from the Oakland shore, was identified as the most vulnerable, susceptible to the greatest seismic loading, and it is the first to go. The whole double-decked section between E-17 and E-18, nearly as long as a soccer field, fulcrums into the bay. The easternmost end remains balanced on the E-18 pier, but the western section rattles off E-17 with an unearthly sound and goes in, causing a terrific splash. It lies in the water like a child's slide. The water is only 10 feet deep, but that's deep enough.

The upper-deck traffic going west toward San Francisco is moving faster and the collapse is so sudden that a couple of dozen vehicles at the cusp of E-17 can't stop and go straight down the slide into the bay. Behind them, drivers brake so hard that their cars skitter and slide across lanes, another mass bang-up that stops traffic dead on the upper deck. Meanwhile, down below on the eastbound deck, the world falls out from under the vehicles on the E-17–E-18 section. Some are launched into the water as if by a slingshot. (Oddly, the two decks remain separated; the upper deck doesn't crush the lower one; they both go down in tandem.) A few cars stay on the roadway, now plunging at a forty-degree angle. Their drivers furiously spin wheels as they gun their engines to get back up the drastic grade. But their screeching tires can't grab the pavement, and in veils of blue smoke they slide down the deck section as if it were ice, and drop into the bay. The vehicles in front are bumped by those immediately behind them, forcing them into the water. They float briefly and disappear.

Seconds after E-17–E-18 goes in, other truss sections fall like dominoes. E-22–E-23. Then E-20–E-21, followed, finally, by E-19–E-20. None of the deck sections drops in flat—they all remain hinged at one end to a pier top. E-22–E-23, the section nearest the Oakland shore, was on the lowest piers, so it drops at the least drastic angle; some cars stay on it, as if parked on a steep San Francisco hill. But dozens of vehicles are submerged.

A few of their occupants, yelling for help, manage to open doors or crank down windows and hurl themselves into the water, which is about 52 degrees Fahrenheit at this time of year. But no one emerges from a number of the cars—the people inside are already unconscious, or they can't get their doors open or windows down. Looking stricken, passengers from cars that went down on the E-22–E-23 section wade to shore—the water is shallow there. Even from E-17–E-18, the downed section farthest out, the Oakland shore is only half a mile away, but the water is ten feet deep. A few swimmers are visible now. Inexplicably, some are heading off for the Port of Oakland's huge north wharf, which is nearly a mile off. Apparently, they can't see that the immediate shoreline is closer. Others, fully clothed, kicking off shoes, paddle toward the concrete bridge-pier supports, which are slick with algae, but where they might find a nubbin to hang on to. The boat with the fishermen picks up eight survivors and makes a drastically overloaded run for Treasure Island's marina. People scream at them as they are left behind. Workers on top of one of the piers being assembled for the new bridge stare dumbly.

Meanwhile, west of Yerba Buena Island, on the heaving suspension span, the shaking has snapped a few of the suspension cables, which continue to writhe like angry snakes even after the ground motion dissipates. On that 9,000-foot section, counting both decks, are probably six to seven hundred vehicles. Gradually, as eastbound traffic backs up through the Yerba Buena tunnel from the first downed truss section, the lower deck freezes. But traffic keeps moving toward San Francisco on the westbound upper deck; those drivers have no idea what's just happened behind them. Below, on the lower deck, hundreds of drivers are out of their cars, wandering in bewilderment and shock, some sprinting toward San Francisco or the island. Those on the now-isolated eastern sections that still stand look down helplessly at the vehicles sinking in the bay, at people struggling to get out as water fills their cars, at swimmers yelling for help.

The Coast Guard has a station on the east shore of Yerba

Buena Island, and several rescue craft are gunning away from their docks, racing after survivors. One driver on the bridge takes several lengths of thick electrical wire out of his utility truck, ties the ends expertly together, and dangles a jerry-rigged lifeline down to the water. Half a dozen swimmers grab for it, and the man on the bridge can hardly handle them all; finally, he and some others knot the end to a railing. About a hundred fifty vehicles, among which are a tow truck pulling a car, an East Bay Transit bus, and a semi-trailer truck, have gone into the bay. The bus has disappeared. The massive semi fell in shallower water off E-21–E-22 and rests half out of the water like a stranded whale. It has fallen on the driver's side, and as the cab fills with water through the broken passenger window the driver—a woman—flails at the door handle, trying to shove it open. Seeing it's hopeless, she manages to smash out more of the window with a piece of wood. She reaches out, grabs the roof sill, and pulls herself through. Grabbing the exhaust stack—still piping hot—she pulls herself off the cab and onto the trailer, which is sinking slowly into the mud—it's no longer halfway out of the water, but most of the way in. Waves begin to ripple over its top.

The strong shaking stopped four and a half minutes ago. There will be aftershocks, some big enough to bring more structures down. But it has ended.

My office is near the waterfront in Sausalito, which began in the 1850s as an Italian fishing village on Richardson Bay, a serene arm of wind-roiled San Francisco Bay, just inside the Golden Gate. The Marin headlands rise steeply behind the narrow strip of bayfront land where the original town was clustered. In the past few decades, as the art of slope construction advanced, the hills rearing over bayfront Sausalito have become dense with expensive homes; few were there before World War II. Some are set on stilts, others are cantilevered with elaborate bracings. Across the street from my office is a restaurant fronting a small beach. The public deck behind the restaurant, which replaced a smelly fish depot and nearby shipworks, offers a fabulous prospect across the bay, a view encompassing San Francisco, Oakland and Berkeley, Angel Island and Alcatraz, and the Bay Bridge. The Golden Gate Bridge is a mile and a half away, around a jutting point of land; you can't see it. But you can look uphill at the stretch of U.S. 101 just north of the bridge, where the freeway exits the Rainbow Tunnel and glides down the Waldo Grade on its way through the heart of Marin County.

My building, a pair of wooden Victorian houses joined together as office space, was recently brought up to code, and wooden structures shed seismic energy most efficiently. Even so, the building seemed about to separate. During the worst of the shaking, the south wall bulged in and out like a loudspeaker at full volume. Two of my pictures bounced off their hooks. As I was running for the door, my tallest bookshelf toppled—too many big volumes were on the upper shelf. I dove for the opposite wall as a half-ton of books and files buried the floor. My bicycle, parked inside, was knocked onto me by the avalanche. I grappled out from under the bike, clawed my way across the pile, and dashed outside. On the way out the door, I noticed that the power had already failed.

The last tremors are dissipating as I hit the street. My building is at the southern end of Bridgeway, the main boulevard through Sausalito, which goes along the waterfront. Some cars are still moving up and down the street.

Their drivers look dumbfounded. Most cars have pulled over, and small throngs of people stand around in wonderment. Everyone is looking across at San Francisco, perhaps expecting to see some skyscrapers down. I take a quick inventory of the landmarks I have been gazing at for years—the Transamerica Pyramid, the Bank of America building, Coit Tower on Telegraph Hill, Embarcadero Center, a Mediterranean apartment tower on Pacific Heights, the sterile monstrosities built during the sixties on neighboring Russian Hill. All the prominent buildings are standing; in fact, the whole city looks undisturbed from here. But there *must* be lots of damage over there—the shock was too powerful.

Which fault was it?

On the hill above my office, coming down Hurricane Gulch—the fog gallops off the ocean through there—I notice a tremendous landslide spilling quietly over Route 101. It leaves me speechless. It is pushing a couple of $2 million homes that, seconds earlier, were perched over the freeway. All the southbound lanes seem blocked by the slide, which keeps pouring down; the soils are saturated and slippery from the last months' rain. When someone else sees it, she lets out a scream of disbelief, and the throng on the street pivots and gasps in unison. The slide appears to go over the divider and stops in the middle of the northbound lanes. One of the demolished homes is sitting directly in the southbound lanes.

The shaking has generated weirdly skittish waves in Richardson Bay that still race in half-circles like nervous fish. Half a mile down the waterfront, I can see that a famous restaurant built on pilings over the water has half-collapsed; the kitchen and back dining room are afloat.

The unearthly noise of the quake—it really was *noisy*, a

bedlam, an inchoate audible maelstrom—has stopped. The bizarre post-climactic hush, where one heard only car alarms, now fills with a vast cats' chorus of ambulances, police cars, and fire trucks. Municipal sirens, meant to warn of air raids or tsunamis, are clamoring too. Suddenly, a couple of police cruisers, followed by two braying fire trucks, come flying down Bridgeway and head up the grade toward the Golden Gate Bridge. My God! I think: *Did it survive?* It was supposed to, especially after $300 million worth of seismic strengthening a few years back, but something bad must have happened up there. I begin to take mental inventory of our emergency stores at home. . . . *My family! My wife and daughters!* My daughters are both in school in Mill Valley, the next town north. The building was seismically improved, and even though it's alongside a creek, probably on densified alluvial soils, it most likely survived. My wife is at work in downtown San Francisco, on the eighteenth floor of a high-rise—a new one, erected in the late eighties. Was it built on landfill? It's on a half-street just off Montgomery, which was the water's edge in the 1850s.

I run back into my office, ignoring the chaos there, and call my wife. The lines are jammed. I call my daughters' school. The lines are jammed. Then, using my cellular phone, I try calling my wife's cellular phone. The cellular circuits must be overloaded; I can't get through. I scribble a note saying I'm safe and am going after our daughters and fax it to my wife. The machine passes the note through, but the display window indicates a problem in the line. Of course—a fax line is a phone line. I'm guessing the phone system isn't just down; it may be wrecked—there's underground cable everywhere, and ground displacement might pull it apart. I call and call, on both phones, and try the fax again. Nothing.

I rode to work on my Zap Electrocruiser bicycle, which, I'm thinking, is a pretty good post-earthquake mode of transportation. It has a half-horsepower motor powered by a deep-discharge battery to boost you up hills and through wind—you're always running into wind or hills around here. Then I realize that there's no power to recharge the battery once I drain it. I struggle with the heavy bike, dragging it over the books heaped on the floor. When I come back out, a number of drivers have turned on their radios. Power must be out everywhere—PG & E tripped its whole system in parts of San Francisco for almost two days after Loma Prieta while it inspected for gas leaks that live current might ignite—but most radio and television stations have emergency generators, so at least we are getting some news. The anchor on KCBS is saying that the earthquake was a speculative 7-plus with its epicenter along the northern section of the Hayward fault. The station has already spoken with the U.S. Geologic Survey in Menlo Park, down near Stanford University, and their initial estimate is that 80 kilometers of northern fault length slipped. (How did *they* get through?) The station's traffic helicopter has just gone airborne. The police and emergency frequencies are talking of massive damage in downtown Oakland and Berkeley, in Richmond and San Pablo, in San Francisco's Chinatown, in the South of Market and Marina Districts—again!—and along the whole I-880 corridor from Oakland down past Hayward. Much of that area is industrial, built on fill. Parts of the Port of Oakland are badly damaged. It's on fill, too, like the portions of San Francisco—except Chinatown, which is on rock—that seem to have been hammered. Near the Oakland port area, elevated sections of BART track have tilted on their concrete supports, and a whole train has fallen off—a drop of about thirty feet. Major highways have sustained damage everywhere. There are dozens of wrecks. Sound walls protecting subdivisions from traffic noise have fallen across highways.

Then the traffic reporter comes in from over the Bay Bridge.

Yerba Buena Island is seven miles from Sausalito, straight across the water. You can see all of the bridge from here, except a portion of the eastern section obscured by a thin streamer of fog dissipating inside the Gate. We've been looking so intently at San Francisco, thinking of 1906, that none of us noticed through the fog's flimsy veil that the Bay Bridge has developed four great gaps on the Oakland side. We are dumbstruck—literally slack-jawed. The traffic reporter is equally stunned—he has trouble finding his voice. Dozens of cars appear to have gone into the bay, he says. They can see hundreds of vehicles still marooned on the structure, mainly the lower deck. Police and National Guard helicopters, several Coast Guard storm-duty craft, and an armada of pleasure boats are buzzing around the downed sections, rescuing survivors. The reporter sees some bodies being hauled into the Coast Guard boats. Many swimmers have made it to shore; some are being attended to by police and tollkeepers from the plaza. Others don't dare; they're trying to hang on to the mammoth, sheer-walled concrete supports until someone rescues them. Helicopters are methodically hauling people up in harnesses. Two people are coming up right now, says the reporter. One is yelling—he seems exultant. The other one is limp, arms and legs hanging motionless. There are several floaters who haven't been taken off the water yet; they've drifted a few hundred yards on the tide.

This is wrenching. I remember looking for days at the crushed-can ruin of the double-decked Cypress Freeway, the Oakland stretch of 880 that fell during Loma Prieta, from my old house atop Potrero Hill in San Francisco. I knew I was looking at forty-two corpses—corpses inside cars flattened to a thickness of one foot. Sometimes I imagined that I was one of those snuffed-out lives. These people on the Bay Bridge, like those people on the lower deck of the Cypress, were doing something utterly innocuous—driving home on a warm afternoon, catching the World Series on the car radio—and were just poofed out of existence. How long does it take the body to expel the soul?

Another car radio is picking up another station and another traffic helicopter—KPIX. Most of what we will learn in the next hours and days will come from the radio—reporters on the ground will have a hell of a time getting anywhere. With electricity and cable out, most of us can't watch television, anyway. The KPIX copter crew have decided to do a quick flyover down the peninsula and up the East Bay to see how bad things are. The pilot is telling us that regional air space has been cleared for emergency flights only, that incoming commercial jets are being rerouted as far away as Seattle and Vancouver. It's not clear whether a news chopper qualifies as an emergency aircraft, but they'll stay airborne unless ordered to ground.

While telling us this, the KPIX copter, which is based at the north end of Richardson Bay—we saw it lift off a minute or two before—has made a sweep across the Golden Gate, directly over the bridge. In the distance, they can see smoke coming off the Marina District, just as in 1989. The bridge looks intact, according to the reporter, but there was a terrible collision. Three ambulances have loaded some injured, but so many others appear to be hurt that they're being laid in the backs of police cruisers. Police are trying to get all the cars off the bridge, but there's so much wreckage they're having trouble clearing a route. A fire truck is pushing some vehicle sideways with its bumper, just shoving the thing out of the way. There can't be nearly enough ambulances in the Bay Area to handle all the injured—probably not even the severely injured. And how long will it take ambulances to get to hospitals, with streets and highways as chaotic as they must be? And then, once the injured begin flooding the hospitals, there won't be nearly enough beds. We haven't heard yet whether the East Bay hospitals have even survived the quake. Most of them are within three miles of the fault.

The copter has a TV camera aboard, which is sending back live footage. I feel a mordant wish to see it. I keep trying, with no success, to get through on the cell circuits to my wife and my kids' school. The KPIX copter has reached the Marina District. Many, probably most, of the homes and apartments there were strengthened or rebuilt after Loma Prieta, but a bunch of them sit on abysmal soils—the old tidal lagoon that we erased. One apartment house has fallen sideways . . . another . . . flames are shooting right out of a manhole in the street. Two buildings are burning; fire crews are already there. Most of the structures look okay, at least from the air, we hear on the radio.

Now they're approaching downtown. The prominent high-rises that define the skyline show no obvious damage. The older skyscrapers, the Deco and brick-faced neoclassic beauties from the twenties and thirties—Mills Tower, PG & E, the Shell Building . . . from the air, they seem undamaged. So do the Nob Hill hotels—Huntington, Fairmont, the Mark Hopkins. (The Fairmont had just been completed when the 1906 earthquake struck, next to Hopkins's vast, multi-spired mansion; both were gutted; the hotel was rebuilt, but the mansion never was.) Now we hear that there's major damage around the waterfront—a couple of huge wharf buildings have folded, some older buildings on the Embarcadero have dumped piles of bricks and ornamentation in the street. South of Market Street, it's even worse. The east end of one side of Harrison Street, between Third and Fourth—or is it Fourth and Fifth? . . . the buildings toward the east end are standing, and one of them is clearly leaning—but down at the west end four buildings have cavitated, four in a row, completely crushed in. There's another below them now, behind the Bank of America branch at Fourth Street and Brannan. One brick wall stands—the building was four or five stories tall. Another one, about the same height, has dropped its entire front wall into the street. A couple of cars—at least a couple—are mostly buried. A woman is on her knees beside the rubble, pounding the pavement with her fists. Traffic is nearly immobilized, wherever they look; there's a vast amount of rubble blocking most of these streets. Police cruisers and motorcycles are slithering through the chaos, using the sidewalks where they can. The crew in the helicopter can't figure what they're trying to do—clear out emergency routes, probably. PG & E and Pacific Bell trucks, ambulances and fire trucks . . . they're down there, too, but the streets are such a mess that, in places, they can't go anywhere. Cops seem to be holding back traffic and clearing free lanes on one-way streets so emergency vehicles can charge through, going the opposite way.

Two fires are visible nearby. They're both south of Market. One is consuming an apartment building; the other, an industrial space close to the waterfront. If water pressure is lost, some fire departments have fallback pumping systems that can take water out of the bay. There's more smoke coming off the Mission District. Far across the bay, at Oakland or Berkeley, more smoke in the air. The reporter says they're going to head down the peninsula to the San Mateo Bridge, cross over, and fly up directly over the fault trace. There are so many helicopters hovering over the Bay Bridge—Coast Guard, army, KGO, KCBS—that KPIX's will steer clear. "The last thing we need now," says the reporter, "is a helicopter collision." Then they change their minds and make a quick pass-by. "This is . . . this is . . . We've never seen this. . . . I covered the Oakland Hills fire, Loma Prieta, but this . . . this . . . The Bay Bridge is *gone*. Bodies—bodies are floating, we see several from here . . . people are still being rescued, live people, survivors . . . we can't . . . no, we *can* see some of the cars that have gone in, we can make a few of them out, under the waves. . . ."

In front of my Sausalito office, a crowd has gathered around this broadcast. Forty of us, or fifty, standing dumbly, staring at the bridge, listening to this report—I can make out the KPIX copter even from way over here, through a set of binoculars that I remembered in my desk drawer. The copter has turned south and headed for SFO. *What in the world should I be doing now?* Since I can't get through on the phones, I should make a bike run to my daughters' school. I decide to keep phoning for another ten minutes.

The SFO tower warned the helicopter away. The controller said there's some runway damage, but military planes may try to land there, anyway. The reporter saw trucks racing up and down the main runways, stopping to inspect the tarmac. Now the helicopter is well out over the bay, approaching the Oakland Airport. Its runways are a mess; you can see damage from several hundred feet above. The tarmac has separated, slumped. At the north end of the main runway, one side has fallen at least a couple of feet lower than the other. The levee there has slumped badly, and the bay is washing over—no, it's coming through a big crack. A lake is forming near the spot where planes, taking off, usually pull up their wheels. The copter drops for a closer look, but then the Oakland tower shoos it off, too. "I don't know why," grumbles the pilot. "No plane's landing here for a long time."

I remember being told several years earlier by Rich Eisner—then the region's Emergency Services chief—how hellish the initial rescue effort would be. The Bay Area's abject dependence on bridges, airports, and freeways built on soft soils—landfill or old bay-margin alluvial benches—would complicate matters tremendously. "We'll have a hell of a time getting around," he had said. "Getting people to hospitals. Getting critical people *from* hospitals that are destroyed or unsafe to other hospitals. Moving work crews in. Forget about getting to work. Retrofit or not, all the major transportation corridors on the east side of the bay will be disrupted while bridges are checked for damage. Landslides, pavement fissures, collisions, downed overpasses, chemical spills, live voltage—it's gonna be something or other. In some places it could be *all* of the above. On

Highway 24 where the fault crosses it, just south of Berkeley, we could have six or seven feet of pavement shift, big fissures in the roadway, collisions, downed power lines, a substation fire or even a wildfire over the same terrain that burned in '91. Landslides. The Caldecott Tunnel through the hill is apt to be blocked. All at once! We'll see versions of it elsewhere. Both sides of the bay. You know generally what's gonna happen, and you suspect where it will be worst, but you can't be certain where. That limits your response efficiency. You can't keep an ambulance crew waiting at the eastern side of the Bay Bridge for an *earthquake*, the way they keep tow trucks waiting for stalled cars at rush hour. You just can't do that, obviously."

KPIX is now over the San Mateo Bridge, a few miles south of both airports, which are opposite each other at the bay's widest point. The eastern bridge approach, the highway leading to it, is ruined. There appears to be liquefaction along a mile or more of Route 92. From up there, they can see traffic backed beyond the western approach too, seven miles across the bay. Hundreds of vehicles are on the bridge, and all are stopped. There's a breakdown lane, where some drivers have parked their cars and begun walking off. They may be afraid that the bridge will still fall in. Most of it is low on the water, like the Lake Pontchartrain Causeway in Louisiana, and it's been re-engineered; they haven't got much to worry about. A couple of Highway Patrol cruisers have made it to the eastern approach, near the toll plaza. The police seem to be configuring a passable clearance route. The copter is directly overhead, hovering at seven hundred feet, and the damage is noticeable. The pavement is simply torn apart. Another truck crawls in, filled with orange cones. Some of the lanes near the toll plaza seem to be under a couple of feet of water. Either bay dikes have failed nearby or sand and water has been forced up through gaps in the roadway.

The helicopter flies on, over San Leandro and north Hayward, toward the fault trace. These are the East Bay flats, densely settled for miles and miles: modest houses, small yards, endless residential neighborhoods. Hundreds of thousands of structures. The pilot knows just where the fault runs here, about a mile and a quarter shy of the hills. This is where you'd expect apocalyptic damage, dead atop the fault. But it's not as severe as you'd expect—so the traffic reporter says. Quite a few houses—dozens at least, but not thousands—have been pushed off their foundations. You can see it from the air: they're bowed, partially collapsed, and the foundation wall is sticking up *alongside* the house. The reporter says it's like a set of huge jaws closed around the homes and shook them off their foundations. Those homes are goners. The helicopter flies over Fairview Hospital, one of the largest public facilities in the area. Several of its buildings have obvious severe damage. Patients have been hustled outside: many are in beds that were simply carried onto the lawn. The reporter peers through his binoculars and can see that some patients on the lawn are still hooked to catheters and medical machines—one assumes that had they been disconnected, they'd be dead. Electrical cord lies all over the lawns—hospitals have generators. A group of firemen, accompanied by some people dressed in light blue—doctors in scrubs, surgical nurses—are trying to pry a way into a multistory building with at least one pancaked floor.

The reporter is even and cool through all of this; I can't

imagine what he's made of. The pilot is much more emotional. "*God!*" we hear him shout, although it's the reporter who's miked. "Look at that!" A big low building has caved in, about a mile and a half from the fault, down near Hegenberger Road around the Oakland Coliseum. The collapsed building, over which they are now hovering, is a huge discount store of some sort, with many cars in the lot; dozens of people must have been inside. The reporter corrects himself—they *are* still inside. Two sides of the building have folded and a third of the roof has dropped onto, or near, the ground. It's resting on the contents of the store—inanimate and human. The roof remains braced to the walls at one side of the store; the unlucky shoppers were on the other side.

"How do you get people out from under there in time?" the pilot wonders over the whuff-whuff-whuff of the rotors. "That section must weigh a hundred tons."

"It's horrible to think about what's happened to those people," the reporter says. "It's beyond horrible." Now he finally sounds grieved.

I toured this area with Rich Eisner in the summer of 1996 and learned what a "tilt-up" building is. "Discount 'box stores,' big warehouses, a lot of light industrial buildings—they're all constructed the same way," Eisner had said. "As construction techniques go, it's very quick and quite cheap. You just don't get the soundest structure in the world. The worst tilt-ups by far are the ones more than thirty years old. Tilt-up construction has improved since the San Fernando quake, like everything has. But the building codes apply to buildings. They just don't fully accommodate the effects of *soils* during an earthquake. Where have we been putting these discount stores, these light-industry and warehousing complexes? The stores move tons of volume on the cheap—little profit margin, just constant merchandising. A lot of the companies renting space in these anonymous office parks are financially marginal. Translation: none of them can afford high rents. Expensive land means high rents. So they build on the last cheap land in the Bay Area, which is the perimeter of the bay! You might have to install drainage, and you can compact the soil, which reduces liquefaction potential—but this still isn't good ground.

"So once they get a permit to build, they prefabricate sections of building wall on the site. Then they use cranes to tilt the sections up. They do that a few more times, then hoist up and attach a roof to the wall sections and you've got a building. Typically there's a huge open space under the roof with minimum floor-to-ceiling bracing—they don't like pillars and columns in warehouses and discount stores because they impede movement and flexible use of the building's interior space. But too little bracing makes the roof joints vulnerable during strong shaking—which you're going to get at a lot of these sites. Even if walls and ceilings stay up, there's all this heavy merchandise stacked on shelves two or three stories tall. There's a wall or ceiling brace to hold the shelf, but you can't tie down the stuff on the shelf. A pallet of stuff they sell at warehouse stores can weigh hundreds of pounds, and they stack some of these pallets and bind them together. Some of that stuff will shimmy itself off. It might take twenty seconds, but some shoppers—how about old ladies?—still aren't gonna react in time. If the roof doesn't flatten you, the merchandise will."

The news team expected to find the greatest damage right

along the fault trace, but down here, at the bay margin, things seem to be worse. They are now over Edgewater Drive, a low-rise office and industrial strip adjacent to the Oakland Airport. Most of the buildings along Edgewater are tilt-ups erected prior to the revised 1973 Uniform Building Code. "The weakest of the weak, if they have not been retrofitted to current building code," Rich Eisner had said as we drove through here. "This is amazing," the reporter tells us. "Unbelievable. It seems like a tornado swept through here.

"We're making out more fires to our north. We've got a bunch of them."

I have watched live news coverage from Bosnia and Chechnya, watched the explosions that killed people on camera, all while doing something incongruous and mundane—shelling a Dungeness crab, combing my daughters' hair—and remained distant and unmoved. Another war zone. More lunatic ethnic hatred. Remoteness is the thing—television's glass wall. Now I am in what amounts to a war zone. It is home. With so many casualties, it might as well be a battleground. But natural disaster isn't war, and because there's no hate and murder involved, it induces a kind of dreadful fascination. We try and try to usurp it, but the power of nature still awes.

The driver whose radio we have been listening to utters the most plaintively helpless question. "What do you do?"

Millions of people must be asking themselves the same thing. What *do* you do? What *can* you do? First of all . . . where do you go?

Traffic has been flowing up Bridgeway toward the Golden Gate—God knows where they think they're going—and now it's beginning to back up to downtown.

More and more cars begin to turn around and head north. One of the drivers stops and tells us he's just heard that, on top of the accident, the bridge is going to get a thorough inspection before it reopens to traffic. That could take days. Then there's the big landslide blocking all the southbound lanes; it has, indeed, spilled over the median divider and into the outer northbound lane. This driver has also heard that there's another jam at the bottom of the grade, about three miles north of the slide. They didn't mention what caused it. "We're trapped," he says. "There's no way out of here except a boat."

Actually, he's right. Route 101 is the only through route north from the bridge and Sausalito; all the surface streets dead-end into Richardson Bay, or Golden Gate National Seashore, which is roadless. It's got a couple of fire roads, but they don't go anywhere. But I've *got* to get out of here—I've heard enough for now, and I'm desperate to find my daughters. I try the cell phone one last time. *Someone's* getting through, just not me. I get on my bike and tear north through Sausalito. Going up Bridgeway, I notice a lot of obvious damage—masonry house walls with serious cracks, a second waterfront restaurant that's drooping off its pier. Inside a liquor store that I ride past, half the bottles seem to be off the shelves. I glimpse a mammoth pile of broken glass and pool of expensive booze, but the cashier is just standing there, listening to his radio.

A couple more storefronts—older buildings, historic buildings—with severe cracks. I wonder how much historic building stock we'll end up losing. On a declivitous hill that Bridgeway skirts, I notice one cantilevered house . . . another one . . . a couple of others . . . that have slipped. They haven't gone far, a foot or two, but what does it cost to get them back up? How do you even do it? I notice a woman standing right under her back deck, which has tilted to one side; she's underneath tons of structure that could crush her if it came all the way down. I think I hear her sobbing. Several other bicycle riders appear around me. A motorcycle—I wish I'd ridden mine today. There's lots of northbound traffic beside us, crawling along, trying to get out of Sausalito. Then, about a mile before Bridgeway merges into Route 101, I run into the backup that the driver just told us about. I'm now streaking past dozens and dozens of stopped cars, feeling vaguely triumphant. What's going on there? . . . a curve before the highway comes into sight. . . . *Water.* Mud! There's a great *mudflat* . . . a puddle, except it looks more like mud . . . sitting across most of the highway, right where the houseboat community sits, at the shallow north end of Richardson Bay. Did a dike break? There is no dike. Now I see three California Highway Patrol cruisers—how do they *get* to these emergencies so fast?—on either side of the freeway, holding all traffic at bay.

I pedal up to one cruiser and ask one of the two cops if bicycles are being allowed through the water and onto the highway beyond. The bicycle path, which veers off to the right, rides along a berm through the Richardson Bay marshland, and I don't like its looks—it's been walloped by ground shaking, it's full of big cracks.

"You mean you want to ride a bicycle on Route 101?" he asks.

"I've got to get home somehow."

He yells at another officer. "Are we letting bike riders on northbound?"

"Beats me."

"If you don't let people get home by bike, how can anyone get home?"

The cop looks me over carefully. "Be careful right here. It's more than water. The pavement is a mess." I notice that the lower legs of his pants are wet.

I see what's happened. Liquefaction has melted the road base, which has dropped the pavement. The asphalt is macerated. Here, alongside Richardson Bay, the water table must be so high that the strong shaking forced it up onto the road—if you squeeze the bottom of a fat sponge, water's ejected from the top. I walk the bike through slowly. There are pavement cracks wide enough to swallow my foot; in the space between there's dense, grippy mud.

Some sections of roadway are still slumping before my eyes; I stop for a minute and watch a piece of inner lane cavitate several inches. A little more than an hour ago, traffic was flying across here at seventy miles an hour. How long before they manage to remove the broken asphalt, then regrade and repave it? Working around the clock, maybe three days. But this kind of road damage has happened all over the place. Where is the equipment supposed to come from?

Whatever the situation—injuries, fires, crumbled roads, rubble-filled streets—we're going to be practicing triage. Who decides which highway gets repaired first? Opening Route 101, especially northbound, seems like a triple-A priority—it's the *only* way north from the Golden Gate. That means all the commuters who are in San Francisco, or wherever, on the south side of the bridge . . . tens of thousands of commuters . . . can't get home . . . for days! Unless . . . no, they can't cross and head north up the other side of the bay, either, because the Bay Bridge is ruined. They can't try going south and cross on the San Mateo Bridge because the highway approaches are out on both sides. . . . I don't know about the Dumbarton Bridge, the southernmost crossing, but the same thing's probably happened there . . . they can't reach the Richmond–San Rafael Bridge, the northernmost crossing, from either side. There is *no way* across the bay, and there's probably no way *around* the bay, unless you take a tremendous detour to the north or south.

I have an eighteen-foot motorboat in my driveway. Its twenty-five-gallon tank is nearly full. There is regular ferry service from San Francisco to Marin, but I suspect that people are already queueing up for blocks, and the ferries—some of them, anyway—must have been drafted into emergency service. I've just thought of the best way—the *only* way for now—to reunite my family: I've got to get my daughters home, however I can, then launch the boat and make a run to San Francisco to meet their mother, somewhere.

The pavement north of the viaduct—it's survived—that crosses the northern tip of the Richardson Bay is on better soil, and is undamaged. This is my first experience riding a bicycle on an empty expressway. After three vaguely exhilarating miles, I exit and pedal through Mill Valley, the toniest town in Marin, which sits in the shadow of Mount Tamalpais. The town looks scarcely damaged as I ride through; I see a couple of cheaper homes shaken partway off their foundations, and a few stucco buildings with ground-to-roof cracks. I pass Tamalpais High School, where students are mobbing over the lawns, waiting for some way to get home. The school buildings are intact; all primary and secondary schools in California's earthquake zones have had seismic retrofits. Power is out here, too; the stoplights aren't functioning. You don't expect anarchy and looting in a town like Mill Valley, and drivers display extravagant civility, politely waiting to cross intersections, waving me ahead. The whole town seems to be out in the streets, clustered around portable radios. Nearly everyone seems to have a cell phone on his or her ear.

Finally, I get to my daughters' school. I'm told that a jitney bus has just left with both of them aboard.

I can catch up. They will use the commuter sneak route parallel to 101, a hilly road called Camino Alto that goes through upper Mill Valley and back down into Larkspur. It's going to be clogged with cars; if you've made it to here, it's one of three or four routes you can use to get farther north.

It's not yet dusk, but most cars have their lights on. A four-mile ribbon of taillights trails up the grade and disappears over the crest. Pedaling with the motor engaged, I ride faster than the traffic. I recognize the school's minibus about a mile up the hill and knock furiously on the door; my daughters, yelling with joy—as am I—have to convince the driver that I'm not some lunatic. Then I have to persuade him to let me take the bike on board. I hoist it up the three forward steps, dump it, and fall all over the girls.

I'm now surrounded by a dozen schoolchildren who are experiencing the biggest and probably the scariest event in their lives. I have no idea what to say to them. The main thing they all want to know is whether their houses are still standing; their universe is small. I tell them that Mill Valley and Sausalito looked mostly undamaged when I went through, so it's probably no different where they live. They should thank God they don't live in the East Bay. A couple of kids whose parents commute to San Francisco ask me when they—the parents—are going to get home. I'm sure the parents are asking the same question—not is my house standing, but how do I find my kids? I know that my daughters' school has a disaster preparedness plan, and I should know the details, but I don't. The driver tells me that any child whose parents or relatives aren't home, and who doesn't want to stay with a neighbor, is going back to school on the bus. Teachers have volunteered for overnight duty, and one—her husband is a fireman—told the driver that she expects to be there a couple of days, at least. She has reached her husband on a CB, and he told her that four Marin ferries are being used for emergencies; three others are operating. Stranded commuters hoping to get home that way are going to have a long, long wait. (And there's almost no service to the East Bay, which has a far larger commuter population than Marin.) Right now, the ferries are transporting patients from damaged East Bay hospitals—a couple of others besides Fairview have been hit pretty hard—to intact hospitals on the San Francisco peninsula; some critical cases are being shuttled by helicopter. Apparently, a lot of people were killed or injured in hospitals.

I have no idea what these kids have heard about the quake. Do they know the Bay Bridge is down, that dozens of people have drowned? Have they heard that hundreds—at least hundreds—are dead just around downtown Oakland? The driver has been listening to the radio through an earphone; he's afraid of traumatizing the younger ones. But several older kids are badgering him about news; we look at each other and shrug, and the driver turns up the volume.

The KCBS helicopter is reporting from over the Berkeley

campus, where the damage is shocking. Memorial Stadium has been yanked apart—a huge seam now separates two stadium halves, one of which is six feet south of the other. Chunks of exfoliated concrete lie around. Both scoreboards are on their sides. A great fissure rips across the east side of the football field. One glass-walled skybox has collapsed into the bleachers. What if the Cal–Stanford game had been under way, as the World Series was when Loma Prieta hit? The reporter doesn't know the campus, but he does recognize Barrows Hall, which appears to be a total ruin. It was erected in the sixties, not that long ago, but building codes were much weaker then. And architects were infatuated with grotesque design flourishes that perform badly in earthquakes—bombastic cantilevered overhangs, odd-blocked segments, crypto-fascist stuff. The Berkeley campus had some gorgeous old buildings that were unsafe, and others that were relatively new, ugly, and unsafe—the big campus construction boom was in the sixties and seventies, under Governors Brown, Reagan, and Brown. (Reagan didn't like it, but couldn't do much to stop it.) An engineering survey completed in 1997 predicted that the performance, during a strong earthquake, of a quarter of the Berkeley campus buildings would be "very poor." To bring the entire campus up to code was estimated to cost close to a billion dollars. For years, Berkeley has been siphoning a vastly disproportionate share of the UC system's engineering budget, but it was never enough. As of late last year, the retrofit program was about half complete.

Berkeley isn't the only campus that took a hard hit. The station's news anchor, who has miked out the traffic reporter, says she now has reports of major damage at Cal State

University, down in Hayward. There are reports of significant damage from Contra Costa College in San Pablo; several buildings are down and there are some just-confirmed fatalities.

"We're about to get some initial regionwide casualty estimates from OES," says the news anchor.

My youngest daughter asks me what a casualty is. I say it means people who have been injured or might have died.

"People *died* in the earthquake?"

"Yes. They may have died if a building fell in on them, or if they got in a wreck on the freeway, or if they went down on the Bay Bridge."

Her jaw opens. She hasn't heard about the bridge.

The news anchor saves me, in a manner of speaking:

". . . at least eleven hundred fatalities in the Oakland area, and obviously, that is a preliminary estimate. Searchers say they suspect there are hundreds of people in buildings they haven't been able to get into yet. We have one report of fifty to sixty dead on the Berkeley campus, where students were attending class in at least two buildings that sustained major damage. We are told that number is expected to go higher. The Highway Patrol reports that the collision on Interstate 80, at the Berkeley waterfront, caused eleven deaths and an unconfirmed number of injuries. This seems to be the worst traffic accident ever in the Bay Area. There is no casualty estimate yet for the Bay Bridge. The Coast Guard says that over a hundred vehicles are in the water. Bodies—we presume dead people—were seen by our traffic helicopter as they were lifted into boats. Again, there were many people who did survive the bridge collapse, even people whose cars went down. We repeat, these are very preliminary casualty reports from places where major damage was concentrated. There is a great deal of damage elsewhere. We have unconfirmed reports of many casualties in several collapsed buildings in the South of Market area of San Francisco. We just now received the first reports of fatalities in the Marina District. All these reports are not, we repeat, *not* confirmed. We have been told of heavy home damage around Richmond and San Pablo and as far north as Napa, but fatalities and injuries seem to be much higher in the East Bay's urban core. Our science reporter, with whom I just spoke, told me that one explanation for that may be the single-story housing that predominates in the suburbs. He says you're in far greater danger if there are several floors over your head. But we hear that poorer parts of Richmond have been devastated, and fires are breaking out there."

My two daughters are dazed by this. All the kids are. Their expressions are blank, but I know their minds are reeling. They are stricken. We have spoken often of earthquakes, and how to prepare for one, and what it will be like, but my wife and I spared them the subject of death. Even to me, while writing on the subject for a couple of years, death loitered in the abstract. Two thousand, three thousand people expected to die . . . I'd read that prediction dozens of times, and each time it became *more* of an abstraction. . . . If that's all the death there is, I thought, then the *survival* rate will be miraculous for a quake this powerful. The Tangshan quake in China killed maybe 500,000 people—*five hundred thousand people*.

But we now know that more than a thousand people, probably two or three thousand or more, are dead. They are buried under tons of rubble, crushed in cars, expiring in hospitals, drowned under the Bay Bridge. A lot of them won't really be found. Pieces of them, perhaps. Some teeth and bones and a wedding ring. I don't know how to get my daughters to fathom this; I don't know if I want them to. Thousands of people die every day and they don't get upset. How is this different?

As one stunning piece of news after another comes through the radio speaker, I try to visualize what it's like out there. Hospitals that collapsed and killed their own patients; hospitals where patients have to be evacuated and taken to other hospitals, where they compete with thousands of arriving earthquake casualties; doctors flying in from Fresno and Bakersfield—probably from Salt Lake and Denver and Kansas City—performing surgery in a hangar at Travis Air Force Base; rescuers groping through buildings reduced to rubble in downtown Oakland, extracting bodies; people returning to homes with caved-in roofs and fallen exterior walls or homes that look relatively unharmed except that they're four feet off their foundations.

From the top of Mount Tamalpais, though, I bet it all looks pretty much the same.

We have finally made it to the crest of Camino Alto, a few hundred feet above and a couple of miles from the western side of San Pablo Bay; from here, we can look over Route 101 at the Corte Madera Creek estuary. The complicated viaduct, with two levels of interchanges leading on or off, is standing. In fact, we haven't heard much about downed freeway overpasses—that's where a vast amount of retrofit money was spent. But one stretch of the highway, where the pavement is at sea level and on fill, is covered with telltale mud. There is an organized bedlam of flashing lights—police cars, fire trucks, ambulances, you can't tell which from what. Lots of sirens. More pavement collapse there. On the far side of the estuary, San Quentin, California's death row prison, sits impassively on bedrock over the bay.

We reach home, finally, a seven-mile trip, after two hours and fourteen minutes on the bus. My daughters outrace me from the bus shelter, anxious to see what has become of our house. Running through two blocks of San Anselmo's downtown strip, I notice bricks in the street and hunks of plaster and the interiors of stores in disarray; in the antique bookstore two blocks from my house, a few books are still on the shelves. Our house seems intact from the outside. Inside, pictures have shaken off hooks, the bookshelves are partially spilled, and the refrigerator has shimmied several inches out of its nook. There is running water, but no electricity or gas. We find our dog trembling under the back porch; I finally get him out with a chicken leg.

While constantly dialing my wife's cell phone, and getting the stock nonresponse, I flick on a radio, grateful that I have laid in a huge supply of batteries. I also have four filled propane tanks standing behind the garage—and a half-cord of firewood, if it should come to that. Washington, announces the radio anchor, is flying out to save us: the head of the Federal Emergency Management Administration is already airborne aboard *Air Force Two,* along with the vice-president, both of our senators, several members of Congress, and a cluster of military brass. Dozens of military transport planes are headed in. National Guard troops are being dispatched from all over California. Doctors are flying in voluntarily from many cities; a request is out for surgeons and specialists in physical trauma injury. Except for helicopters and perhaps some STOL aircraft, every incoming plane will be forced to land either at San Jose or at Travis, which is fifty miles northeast of the Bay

Bridge halfway to Sacramento. Liquefaction damage to the commercial-airline runways at Oakland's airport is severe; only helicopters are landing there. But there are two separate runways for private planes, and they came through the quake in better shape; they may be usable in a few days. At San Francisco International, one runway is already being regraded; fixed-winged planes will be able to land there within the week.

The radio suddenly cuts to an interview with California's Office of Emergency Services (OES) regional administrator. He is talking about the issue that seemed to obsess his predecessor, Rich Eisner, when I interviewed him in 1997 and 1999: logistics. Even with emergency personnel and supplies being flown into Travis, he says, getting them to the hardest-hit areas is a huge problem. He delivers another stunning piece of news I haven't heard: the older of the two parallel Carquinez Strait bridges, which take Interstate 80 over the Sacramento River where it debouches into San Pablo Bay, is badly damaged. It is out of service indefinitely. The parallel span is undergoing inspection. If engineers fail to find bolt shearing or strut damage, it may be open soon, but probably just to emergency vehicles at first, and even later a couple of lanes will be reserved for them. That is another epic constriction point—both bridges separate the Travis air base from the Bay Area. The nearest alternative crossing, the I-680 bridge, is a number of miles east.

The OES regional administrator appeals to anyone who owns a large pleasure craft—an ocean cruiser fifty feet in length or longer—to lend it to the emergency effort. He supplies a toll-free number and extension. OES is assembling a makeshift pleasure-boat fleet while negotiations continue with ferry companies around Puget Sound, where they have much larger car-carrying ships. If we can pry some big ferries loose, he says, it will still take several days for them to arrive. With the bridges and approaches so devastated, discussions are under way with ferry operators "worldwide." What is worldwide? asks the reporter on the phone with the harried state official. "Europe, the Philippines, eastern Canada," he replies. "We're not sure at this time of year that some of the ships—even big passenger ferries—are capable of a Pacific or Atlantic crossing. They'd have to be refueled en route. We don't know if the countries can spare any. It's gonna be days or weeks, if at all. We have to make do mainly with what we've got, and we do have lots of big pleasure boats berthed around the bay." Even in the midst of this, the man tries a joke: "The state will pay for gas."

As the coordinator of the state's emergency response to the earthquake, the administrator says he has to get off the air. The reporter on the line asks him to sum up—"bottom line, very briefly"—just how paralyzed we are. Well, he says, the Bay Bridge and one Carquinez Strait bridge are badly damaged. We probably won't have the new Bay Bridge on line for a couple of years even if we waive all the permit requirements. That is going to create the worst and longest commuting nightmare this country has ever seen—the Bay Area's whole highway system is connected in some way or another to the Bay Bridge. The San Mateo and Dumbarton and Richmond–San Rafael bridges all experienced severe liquefaction at one or both approaches, and there may be structural damage as well. With luck, assuming no major structural damage, they will be back in service in a few days. There is extensive liquefaction on Route 37, the highway around San Pablo Bay, and on Interstate 680, the main route around the southern end of San Francisco Bay. The worst liquefaction is on Interstate 80 down at the Berkeley waterfront. The lanes on the bay side will be impassable for weeks. Meanwhile, the BART tunnel through the Oakland hills suffered several feet of displacement, and service to the outer East Bay suburbs is gone, indefinitely. The elevated track near the Port of Oakland has leaned, dropping a train. A long elevated section of track near Hayward has collapsed. There were many injuries and some casualties. BART functions only in San Francisco for now.

In short—the state employee with a latent penchant for career suicide tries a somewhat risky medical metaphor—we are somewhere between paraplegic and quadriplegic. For the next few days, anyone trying to get from one side of the bay to the other will face immensely involved detours. Routes around the bay, and the other bridges, when they are back in service, will be jammed for longer than most people want to think. We will be using roads and streets designed to carry a minor fraction of the traffic routinely handled by the freeways; we will have—presumably—three trans-bay bridges doing the work of five. But that understates the problem, he says, because the Bay Bridge carried as much traffic as the San Mateo, Richmond, and Dumbarton—and the Golden Gate. The Golden Gate Bridge sustained what seems to be minor damage, the great collision notwithstanding.

Major employers are being asked by the governor to discourage all but the most strategic workers from trying to commute in the days and weeks ahead. Sudden street and road closures will be the norm, anyway, especially with the expected aftereffects.

The reporter won't let the state official go. "What sorts of aftereffects?"

"Well, the sewer system in Oakland and Berkeley and elsewhere has probably sustained a lot of damage. It's only a matter of time before raw sewage starts backing up into the streets. I'm sure it already has. We're laying down a makeshift bayside interceptor system to replace the main one, which got whacked pretty badly—it picks up almost everyone's outfall in the northern East Bay. That's going to cause a horrible pollution problem in the bay, because that stuff is going to flow in there through the bypass drains. The main treatment plants have all gone into bypass mode. The *good* news for the Bay—and I hate to put it this way—is that, in the East Bay, we expect 70 percent of the water service west of the fault to be lost for several days. There is just three days' supply in the reservoirs west of the hills, and much of that will be lost through broken and leaking pipes. A lot of people will be out of water within a day. We don't yet know which trunk mains survived where they make the fault crossing, but be prepared that many didn't. East Bay Municipal Utility District is out there making repairs already—this could become a critical issue in fire-fighting operations. Everyone east of the fault and east of the hills should be in reasonably good shape."

"Why is that good news for the bay?"

"Because as neighborhoods progressively lose water pressure, you won't be able to flush your toilet. We'll have Port-O-sans in a lot of these neighborhoods soon.

"For the most part," the regional coordinator sums up, "what we expected to happen, happened."

God Almighty. Flick the dial. KPIX helicopter. The reporter cannot stop talking about the traffic, which is unspeakable—infinitely worse than the worst rainy-season, collision-filled rush hour he has ever witnessed. If any freeway moves—any section of freeway—it moves barely. It will just get worse as the downtowns and malls continue to empty out. A lot of fire trucks and ambulances seem to be moving from San Francisco and the peninsula toward the East Bay, taking 101 past the San Francisco Airport down to the Dumbarton Bridge, where Caltrans seems already to have engineered a route through, or around, the liquefied reach of Route 84 at the western approach. Route 101 south of San Francisco is turning into a long caravan of flashing lights and wailing sirens moving down the breakdown lane, alongside twenty miles of stalled traffic pointed south. Motorcycle police are blocking everyone who tries to break out of the jam and sneak by the same way. The traffic reporter remarks how few people have tried to sneak into the breakdown lanes—a sign, to him, that civilization hasn't broken down yet.

I buzz compulsively from station to station, wishing morbidly that I had a generator so I could make my own electricity and see all this on television, not just visualize it. Along the northern end of San Pablo Bay, Route 37 is flooded for at least two miles, and more of it is going underwater. The dikes at the bay margin—erected long ago to permit marshland to be transformed into pasture—have failed, and in places the incoming tide is flooding over the highway.

Many of these reports must be reaching the stations by cellular phone. A reporter for the *San Francisco Chronicle* once told me that, during Loma Prieta, he was at Candlestick Park watching the World Series game with a couple of colleagues. They went from row to row yelling for anyone who had a cell phone and finally found one and conscripted it into service. Fifteen years later, though, everyone in California seems to have a cellular phone; I have watched Hong Kong businessmen walk down the street speaking into two cell phones at once. I still haven't reached my wife.

Finally, I get through. She's stuck in San Francisco, marooned in her car on Pacific Heights, near the Doyle Drive approach to the Golden Gate Bridge. No northbound traffic is being allowed on because of the inspection taking place, not to mention the ruined reaches of 101 at Richardson Bay and Corte Madera. She headed for the bridge out of habit, and now has no idea what to do. Traffic below her on Lombard Street, which leads into Doyle, is dense in both directions—everyone is being rerouted around, and ushered out of, the Marina District as the fires there spread. She considered leaving the car and walking all the way back to the Ferry Building to catch a boat to Marin, but then she heard that the waiting lines are many hours long. So she is sitting in her car at the corner of Divisadero and Union, utterly disconsolate, watching a block of the Marina burn up. Three miles to the west, the Golden Gate Bridge is eerily empty. Only a few service vehicles and police cars are on the deck. All the wrecked cars have been cleared, all the stalled traffic released, and there's no grossly visible damage—no severed cables, the deck's not out of shape—but no one is being let on. She can't figure why there's so much traffic down on Lombard. Either those drivers are deranged optimists—the ones aimed toward the bridge—or they literally can't extricate themselves, with the Marina burning and so many side streets blocked off. She's not close enough to see.

I tell her to forget about the ferries. I'll leave the girls with our next-door neighbors, launch our boat, and pick her up in a couple of hours at the old Coast Guard dock at the Presidio.

A 175-horsepower car goes twenty to thirty miles on a gallon of gas; a 175-horsepower motorboat, which has no rolling inertia, burns a gallon every four or five miles. A run to San Francisco at top speed could use up ten gallons of gas—nearly half the tank's capacity. I should save it as our emergency fuel supply. When power fails, stations can't pump gas from their underground tanks. Even after electricity is restored, all the supply routes damaged or blocked, it will be hell finding any gas.

After speaking briefly with our daughters, my wife has to tell me more about her ordeal. She fled her building—which swayed several feet back and forth—down the stairs, along with everyone else on twenty-four floors. The streets were mobbed with tens of thousands of people evacuating several dozen high-rises at once. They ran everywhere and froze traffic dead. Everyone's worst fear was glass falling off skyscrapers, but there was none of that, at least not on her street. Then it took her an hour to get out of the parking garage. The fumes were so bad underground that attendants were running all over the place, ordering people to turn off their engines until they could move three or four car lengths; then they were told to shut them off again. It was claustrophobic and tense down there; some drivers were so convinced of their imminent doom—from fumes, aftershocks, who knows?—that they turned back into parking stalls and fled up the ramp on foot. When she finally got to the street and began snaking through hideous traffic, her usual route home was blocked: a couple of streets had been dedicated as emergency corridors. Traffic was detoured this way and that: she had to zig and zag around China Basin, south of Market Street, near the Giants' new stadium—which looked completely unfazed; it's on shaky ground but was exceptionally well built. All the South of Market streets had stretches of deformed pavement, and here and there she could see a great deal of damage. The worst was down Sixth Street, where lots of semi-indigent people live in rotting old hotels and subsist on pastrami, crack, heroin, and alcohol. A couple of the SRO hotels had crumbled or shed outer walls; there were several fires, piles of debris in the streets, and ambulances everywhere. She had to detour over Potrero Hill and up through Haight-Ashbury to outer Pacific Heights—where she has just arrived. It has been three hours and forty minutes since she left her office.

Before breaking the connection, I tell her I'll try to call from the boat when I get close. If I can't reach her, and if it's dark by the time I get there—it will be—I'll be tooting the air horn and waving my arms.

It's February, near the end of the rainy season. Thankfully, it's supposed to rain tomorrow; that will help with the fires. A pair of storms, tracking slowly across the northern Pacific, have produced a heavy ocean swell; the waves are marching through the Golden Gate and dissipating in the bay, but I'll be making a run quite close to the Gate, and it may be rough. It's foolish to take the girls, and they don't want to come anyway. I park them with our neighbors, who are serving barbecue by candlelight while glued to the radio. They've just heard a shocking report: terrible damage in the Canal District, a poor section of San Rafael close to where we live and exactly where I'm about to go through to launch the boat. I quickly hitch on the trailer and head over there.

The trip takes longer than usual because all the stoplights are out, and there are twenty-three between my house and the marina at Loch Lomond. When I pass under Route 101, all southbound traffic is being led off; I assume it's because of the liquefied roadbed at Corte Madera and Richardson Bay. There's no northbound traffic at all, except for a lone motorcyclist who must have sneaked on somehow. At an intersection, I hear a man telling two of his friends that the sound-wall along 101 as it passes through San Rafael has fallen across the roadway; they're clearing it now, but it's a great mess. It's just north of where I am. The San Rafael Canal, a navigation channel that pokes into downtown from the bay, comes up ahead. Just across it is the Canal District, densely populated by Salvadorans, Mexicans, and other Latinos who serve Marin's affluent class.

It's bedlam over there—fire trucks, ambulances, police cruisers, and two large cranes at work. I have just enough dwindling twilight to make out what's going on. Two big apartment houses along the canal are down—pancaked from three or four stories to one or two.

My memory regurgitates the most horrifying images from Kobe and Mexico City: apartments and office buildings without a fourth floor, or a fifth. The same thing must have happened here, for the same reasons: shoddy construction on jellylike soils. All of the Canal District is atop old bay muds or landfill. The thought of dozens of people, maybe even hundreds, trapped inside a squashed building—a building you're staring at—induces the sort of delayed shock you feel after you've seen a dog smashed by a speeding car. For a moment I can no longer drive, so I pull over and watch. One of the collapsed buildings is three hundred yards away, directly on the canal. I've cruised by it a few times in my boat and wondered how it would perform in an earthquake; the bottom floor, a parking area, as I recall, was a huge open space with feeble-looking concrete stilts that held up the three floors above. The structure was pre–San Fernando, at least thirty-five years old. Dismally cheesy, it was the sort of apartment complex that communism built in Bulgaria. A mob of ambulances encircles it; I assume they're waiting for rescuers to pull people out. Those must have been the sirens we'd been hearing all afternoon.

The police are keeping a crowd at bay—family, friends, relatives, neighbors. Our babysitter lived in the Canal District for a while, when she first emigrated from El Salvador. She and her husband had to work full-time, with some help from their eldest child, just to pay the rent on that miserable flat. Multi-income families are common over there, and financial need may have saved a lot of lives: at three in the afternoon, most of the tenants were probably at work, and older children should still have been in school. But even ten-dollar-an-hour nannies have to hire five-dollar-an-hour babysitters to watch their own toddlers.

In that crowd, I suddenly realize, there are shattered parents who were at work while their children were at home.

Without quite realizing it, I have begun to sob. This is the worst of the awful. It is gut-wrenching; it is something beyond that. Right now I don't want to look. I drive on to the marina, launch the boat, and aim in the general direction of Alcatraz, whose lighthouse is casting its startling beam. With the throttle fully open I can do almost forty miles an hour. I have running lights but no headlamp;

everyone else had better have running lights, too. Right away, I see a number of pleasure boats racing across the bay. As I approach the city, the bay is jumping even more with boats; it could be a Fleet Week Sunday. There are ferries, tugboats, Coast Guard rescue craft, a navy ship of some kind, big cruisers, speedboats, charter salmon boats, Zodiac rafts, even a couple of commercial fishing boats. Wherever they're going—they seem to be going everywhere—they're going as fast as they can.

I call my wife's cell number and, to my amazement, I reach her. The motor is so loud I can't understand a word she says. I yell that I'm just east of Angel Island, and hang up. Then I leave the island's protection and meet the ocean swells that have squeezed through the Gate. They are alarmingly high, but quite broad and harmless, rolling languidly toward Oakland; the boat rides them like a car going over dips. On any other night this would have been a glorious run.

There were moments, as I was contemplating this idea, when it seemed just marginally sane. But it has worked perfectly: my wife is there on the old Coast Guard dock, amid a small throng of people who, she tells me later, are all waiting to be picked up. In ninety seconds we're racing back to San Rafael.

At the dock, my wife had been listening to a live report from the Canal District. Not two but *four* large apartment complexes sustained dismal damage. Each has dozens of units, and hundreds of people are presumed inside. We both know that many of them are dead, but we keep that knowledge to ourselves.

I counted several fires on my run to the Golden Gate—a vague glow emanating from somewhere in darkened South of Market or the East Bay, a column of smoke and sparks off the Marina District, dissolving into the night. Now, as we head back to Marin, we notice that one fire—I'm not sure I even saw it earlier, since it was behind me—is awesomely large. It's across the bay from San Rafael, in Richmond, not far north of the Richmond–San Rafael Bridge. We see it roaring behind some small hills as we speed under the cantilever span of the empty bridge, which has workers with floodlights examining the bracing. It's the big Chevron refinery, or some chemical plant, or perhaps a railroad siding where they load tank cars. Every now and then there's an explosive burst and a vast expurgation of dense smoke. Other than the Malibu fire and a couple of similar wildland infernos, it's the biggest fire I've ever seen.

There's so much going on it's almost stupefying. Another incidental cataclysm one can do nothing about. My wife doesn't want to see the Canal District horror, so we crank the boat onto the trailer and take a detour home. On the way, the governor makes her first address. A humorless, tooth-clenching woman to begin with, she sounds as somber as Lincoln at Gettysburg. She runs through the high points: Many people died in the collapse of the Bay Bridge; authorities still don't have an accurate count, and, of course, names are not yet being released. The chain-reaction collisions—there were several—along the Oakland, Berkeley, and Richmond waterfronts killed sixteen and injured many more. The destruction of one old hotel in Oakland caused sixty-six confirmed deaths, and many more casualties occurred in older apartments and SROs that imploded catastrophically; it will take days to dig everyone out. (Where are we going to put all the *bodies* before burial? I wonder.) A number of people whose homes were in the violently shaken corridor within a few hundred feet of the fault were killed, with many more injured, but the heaviest casualties appear to be in downtown Oakland and Berkeley, on the Berkeley campus, South of Market in San Francisco, the Canal District, and off the Bay Bridge. Crews have been so overwhelmed that they're just getting to the landslide areas in the East Bay hills north of UC Berkeley, where many residents, if they were at home, appear to have evacuated in time. But some structures are partially buried and there may be survivors inside.

(The landslides in the East Bay hills are more stunning news that I hadn't heard. Those are the same hillsides scorched by the Oakland wildfire in 1991. I wonder how many people who replaced their burned-up half-million-dollar homes with insurance-financed million-dollar homes have now lost those, too.)

Other areas hit very hard, the governor goes on, were the Oakland Inner Harbor, the vicinity of Lake Merritt, where at least one high-rise apartment building suffered a partial collapse, the Berkeley-to-Richmond waterfront, and Treasure Island. Every hospital in the Bay Area is overfilled, with surgeons operating in corridors; two in the East Bay are still being evacuated due to damage and suspected structural instability. (I think: How long does it take to get a single patient in intensive care, hooked to several machines and catheters, out of a hospital? And then to where?)

The governor tells us the obvious: The Bay Area is paralyzed, prostrate, and will remain so for many days. All lifelines have been devastated; most of the region has lost either power or water or gas or sewage outflow; some neighborhoods, mainly west of the fault in the East Bay, have lost them all. All major highways have impassable stretches. The Bay Bridge collapse is the worst problem; it may be repaired even as the new parallel bridge is built, but that crossing will be out of service for many months. The Richmond–San Rafael and San Mateo Bridges appear lightly damaged, but the approaches—on both sides in the case of the San Mateo, on the west side of the Richmond–San Rafael—are disrupted. Caltrans expects to restore access within one to three weeks, provided the bridges themselves are safe. The Dumbarton Bridge has already opened to emergency vehicles; it's the only way across. BART's Concord Line tunnel has sheared and may be out of service for some time; elsewhere there is damage to many sections of track, especially the elevated stretch where the train went off. Many injuries in that mishap (she actually calls it a "mishap"), but most survived. Fortunately, the train wasn't full. The railroad tracks running along the margin of the East Bay are heavily damaged; it will take weeks to repair them. Shipments off-loaded at the Port of Oakland can't be moved by rail or road, and incoming ships will have to turn back for other ports. Restoring all these services will be a mammoth, expensive, and time-consuming effort.

Having announced all this, the governor slips into her most comfortable mode—a fierce, jut-jawed can-doism, talking about our strength and uniqueness and how Californians can deal with every sort of adversity. The gold rush spirit, and all that.

I can't handle it. We're no better here than anyplace else.

Some friends a few blocks away are ultra-prepared—they have a generator and a hilltop television antenna to back up their cable reception, which has failed where we live, along with the power. We gather the girls and a donation of some wine and a few batteries, and walk to their house. About forty other friends and neighbors have had the same idea.

My imagination couldn't quite conjure these images. Helicopter footage taken before nightfall shows parts of the Berkeley campus disastrously undone. Memorial Stadium looks like a train plowed through the middle of it. Several older buildings are brickheaps. A couple of big newer academic halls, postwar concrete shear-wall structures, are also badly damaged—still standing, but with roofs cavitated and decks lopped off and porticos crushed. Then there is footage of the Port of Oakland, nearly all of which was built on fill. Lateral displacement and settlement have canted a huge dock at a crazy angle and toppled a concrete seawall. Great lifting cranes lie horizontal in the water. Containers are piles of fallen dominoes. A second long length of seawall is down. The station's science reporter says that the strongest ground acceleration anywhere was recorded at an instrumented building near the port—0.81 g. Officials have announced that one wing of the port may still be serviceable, but much of this great facility—the mainstay of Oakland's economy—will be down for months.

Now we get to see the landslides.

With the bridges out, Channel 2, KTVU, which is based at Jack London Square in Oakland, has an advantage over the San Francisco network affiliates: it has its mobile equipment in the East Bay. One of its reporters is broadcasting from a television van in Kensington, the upscale older suburb just north of Berkeley, and the camera is slowly panning a section of hillside where about seven homes are . . . moving. They are sliding downhill in a slow and stately manner.

No—the homes aren't sliding downhill . . . the *hill* is sliding downhill. All of us clustered around the television are raptly speechless. The whole hillside is moving! The soundman trains a mike on the scene, and we hear an occasional bizarre sound—a wrenching, a popping, a slow agonized creaking—and then tiny torrents of water shoot out of the hillside. Everything connected to the transient houses is severing—water hookups, sewer lines, power cables and lines, foundation and deck pilings. As the hillside oozes downward, some homes surf it seemingly intact. Others begin to crack apart—walls bend inward, roof joints separate, a chimney is left behind. These homes, which are worth a million or more each, are disintegrating before our eyes. A fire truck comes huffing up an adjacent street, siren screaming. Gas leaks! Even if the main distribution valves have been shut off nearby, there is residual gas in the lines that could ignite with one spark—and the friction of a big structure sliding downhill while breaking apart could easily create sparks. The television van has only a small spotlight, and can't illuminate this whole spectacle at once. There is no moon, no background glow of electrified urban civilization—it is pitch black where the spotlight doesn't focus. The descending homes are dark angular shadows, moving weirdly, improbably, emitting strange sporadic noise; then, suddenly, the spotlight hits one of them and we can see it breaking up.

Abruptly, the hillside stops moving. The slide has moved only twenty feet or so, but all these homes are goners. Bill Lettis, the consulting geologist, when I spoke with him, had talked about the potential for *hundreds* of deep-seated landslides in the East Bay hills. I suspect this is one of many.

More landslides: Someone in San Leandro, just south of

Oakland, caught a slide above his home with his video camera and managed to bring the tape—by motorcycle, the station anchor said—down to Jack London Square. This slide behaved differently—it dropped downhill at such a speed that one house rolled over. Channel 2 cuts back to the reporter at the Kensington slide, but prefers the action caught on the home video, so they cut right back to the station anchor and run the San Leandro video twice more, in slow motion this time, with the science reporter, who knows something about slides, explaining the difference between shallow slides of highly disrupted material—more like a rock or scree slide—and the deep slump movement that probably caused both of these landslides.

We see some live footage of sewage overflowing into a Berkeley street and police rerouting traffic. Then back to collisions, footage taken earlier. Do police and insurance adjusters have to come out and examine each collision, and file diagrammatic reports, before the vehicles are cleared away? How can anyone even think about that now? Imagine the claims battles, with insurance lawyers arguing that everything was the earthquake's fault—an act of God, sorry, not covered in your policy—and lawyers representing people who were hit by other drivers insisting that, despite the ground motion, the other driver should have been able to control his car. Do we know who hit whom? Who's keeping records? Sooner or later, people will turn litigious—who's more litigious than Californians?—and this earthquake will capture California's courts for years and years.

Cut to a helicopter hovering at a safe height above a refinery fire at Richmond. It must be the one that we saw from the boat. There have been reports of dozens of fires, and many are still burning, but the wildlands haven't ignited—it is too wet—and no major fires have converged along a front. This refinery fire is so intense we can hardly see what's burning; a cracking tower seems to stick out of it. Spasms of smoke roil away like thunderclouds. Big sections of Richmond, we are told, are being evacuated—not just due to the fire and noxious smoke plumes, but due to suspected chemical leaks and the simple fact that hundreds of flimsy prewar homes are damaged. They are uninhabitable, red-zoned, and residents who insist that they *are* still habitable are being removed, if necessary, by police escort. Estimates of homeless have already climbed into the tens of thousands.

The burgeoning numbers of homeless have stolen center stage, for a while. Apparently, even some schools that were extensively re-engineered and designated as post-quake shelters have sustained damage; a couple, in El Cerrito or San Pablo, were directly on the fault, and the shaking was so strong they fell apart. There are no reports of casualties, but almost all students were out of both buildings when the quake hit. Some kind of half-day holiday. (Finally, a miracle.) Meanwhile, intact school and campus buildings in the East Bay are now being converted to shelters, and several are already filled; as more homeless pour in they will have to be relocated out of the Bay Area: to Sacramento, to Stockton, to the old Fort Ord base near Monterey. Mary Comerio, a professor of architecture and planning at Berkeley, told me that many people rendered homeless by the Northridge quake refused to go to shelters; they parked their cars outside the remains of their homes and slept in them for weeks. They didn't want their property looted, and would rather sleep in a familiar car than somewhere strange.

After leaving our friends and putting the girls to bed, my

wife and I had been sitting on the deck with the radio, watching the glow of the Richmond refinery fire jaundice the sky. At four o'clock in the morning, a couple of Ativan and another slug of wine sink me into a brief, troubled sleep. I awaken at seven, struggling out of a morbid dream, and as my brain clears a shock of remembrance hits. In a second, I am bolt upright. We still have no electricity—the blank face of my alarm clock looks impassively at me. We heard last night that a couple of 220-kilovolt substations went down; one of them sends power to most of Marin. Three-wired transmission circuits handling such mega-voltage need to be widely spaced, and the tall, long insulators connecting them to transformers are brittle, highly susceptible to ground motion. Transformers are grouped in banks of three, and if one fails, so do the other two. There were no predictions as to when power might be restored here, but some parts of San Francisco have gotten it back.

I turn on the radio to a report of mass flooding in the Delta.

This is a breathtaking piece of news. I pick up the phone and try to call our friends with the generator, but there's no dial tone. I recall that these particular friends don't have a cellular phone—which is inconceivable, almost *inexcusable,* in Marin County. I scribble a note for my wife, who lies oblivious, and run back up the hill to their house.

They are alone now, but still gathered around the television. The images were so mesmerizing they couldn't drag themselves off to bed, so they spent the night on their living room couch, dozing intermittently. For the past hour, they tell me, the unfolding Delta disaster has wiped most other coverage off the air, except for summaries every few minutes of the quake's most tragic and paralyzing effects. A hundred thirteen people are confirmed dead in the collapse of the east end of the Bay Bridge, and the count continues to rise; divers have been recovering bodies from submerged cars all night, and they believe there are more down there. In the greater Bay Area, more than two thousand are now confirmed dead. Eighty percent of the casualties were in Oakland and the older East Bay suburbs, but luckless people died far away due to some freakish related events: rockfalls onto a Santa Cruz beach, merchandise sliding off store shelves up in Napa. Structural damage alone—I don't know how they come up with these figures so soon—is estimated at $35 billion so far. Productivity loss will cost tens of billions more.

While I absorb that gruesome recitation, the television is showing some mesmerizing footage shot from a helicopter over the Delta. Its gridwork of farmed fields inside huge levee-ringed tracts of land has transmogrified into a vast sweep of water, miles and miles across. The crests of levees that for decades prevented such inundation, the top couple of feet or so, are still visible, weaving sinuously through this giant newborn sea. With most of their ponderous mass hidden beneath the water, enormous levees look like the little check dikes that rice farmers sculpt up in their field. Here and there a levee crest line just disappears, for some distance—two hundred, three hundred feet? Those are the breaches where the levees failed and the surrounding water roared in, flooding these sub-sea-level depressions Californians have always referred to as "Delta islands." If we'd called them "unfilled reservoirs"—exactly what they are, or were—we might have abandoned the expensive effort to hold off the inevitable.

The cameraman in the copter zooms in on one breach

after another; he brings back the focus to show a levee crest-line forming one whole side of a former "island"—it's four or five miles long—that has been breached in three different spots.

A Department of Water Resources spokesperson is in the helicopter with the pilot, reporter, and cameraman. Few Californians know anything about the Delta and its exquisite vulnerability, so this mass inundation must be a shocker. It's probably blindsided the media, too; most reporters I know never quite understood the hydrodynamics of the Delta region or why it needed to be "fixed," but never was, at least not enough. The DWR spokeswoman is a hydrologist who has worked in the Delta for years; her bosses must have pushed the poor woman out to make sense of all this to a bewildered public and media.

Sherman Island, the westernmost of the leveed tracts, and the one nearest the epicenter, went first. The north-side levee failed in two places. That reach of levee runs right along the main Sacramento River—not some slough or side-channel—and with the Sacramento carrying most of the Sierra Nevada's runoff at that point, about 90,000 cubic feet per second right now, the entire island, some 10,000 acres, was filled to sea level in a couple of hours. Route 160, one of the three highways crisscrossing the region, runs atop the north Sherman Island levee for miles. In fact, Route 160 is on one levee or another all the way from Antioch, at the western edge of the Delta, to Sacramento, fifty or sixty miles northeast. The highway is now gone, erased, in two places along Sherman Island; it's impassable elsewhere due to levee slumps and cavitated pavement. Route 4, the other east–west Delta highway, has liquefaction damage all over the place, but may still be marginally drivable. It's closed to all traffic except emergency vehicles and evacuees. A mass evacuation of nearly all Delta residents—the few thousand people who inhabit its half-dozen towns, and dozens of farm families living out on an island somewhere—has gone on all night. A lot of people were taken out by boat, leaving everything behind. Fortunately, thousands of pleasure craft are docked at a number of Delta marinas, and many boat owners leave a duplicate set of keys in a marina safe, which is how a vast rescue fleet was assembled in a hurry, racing up and down rivers and sloughs picking frightened people off levee tops. Despite this amazing evacuation, a number of people are reported missing and are believed to have drowned; houses near levee breaches were broken off their foundations and swept toward the centers of islands by the inrushing, freezing water. A number of survivors were picked off the roofs of floating homes and trailers by helicopter. They were lucky, because most of the dozens of helicopters that have flown into the region have been busy ferrying patients and doctors to hospitals. But a pair of navy copters was over Suisun Bay at the time, en route from Travis to Oakland. When they saw the levees begin to fail, they knew there was rescue work out there.

The DWR woman says that Twitchell Island, Webb Tract, and Jersey Island were initially inundated within minutes of Brannan Island; their levees probably failed at about the same time, but, since they're farther from the fault, the breaches probably weren't as big or deep; it took the surrounding channels a few minutes to cut all the way through. Then Jersey Island and Bethel Island began to go. All the breaches were caused by slumping and cracking, produced by liquefaction of underlying soils. The whole west-central Delta region, where the subsidence was deepest and the islands are closest to the Hayward fault, is under ten to twenty-two feet of water. It has become, overnight, a vast new inland arm of the sea—an 80,000-acre extension of San Francisco Bay.

Department of Conservation inmate crews, assisted by locals roused in the dead of night, made a heroic effort to save Webb Tract when the initial breach was still just a few feet across. But the San Joaquin River's main channel, pouring into the break, eroded the breach faster than hand crews and a couple of bulldozers could stopper it. A barge filled with rocks for a riprapping project was anchored a couple of miles away, but its tugboat was somewhere else. A pair of cabin cruisers tried to tow it to the breach, but the barge was so heavy it snapped all their ropes. (It wouldn't have helped, anyway; you can't stop a river with a pile of rocks.) The breach was ultimately two hundred yards across, and as the San Joaquin poured in it ate thirty vertical feet of levee wall down to its two-hundred-foot-wide base in half an hour.

Mandeville Island is going under now! The vast open fetch of water across six adjacent, flooded islands—Webb Tract, Frank's Tract, Sherman Island, Twitchell and Jersey and Bethel—built waves that slopped over and melted their own east-facing levees, exposing Mandeville's opposing levee side to intense wave action for several hours now. (A high-pressure ridge anchored over the Great Basin has churned up an insistent north wind, which began to blow around midnight, eventually raking the fetch with four-foot composite waves that Mandeville's quake-weakened levee couldn't repulse.) The same awe-inspiring inundation that drowned the other six islands is now occurring live on TV!

From a thousand feet overhead, the biggest of the Mandeville levee breaks—there appear to be two, and another just forming now—looks like an Amazonian waterfall: wide and low, a roaring muddy tumult. A couple of dozen workers from a state inmate crew are on either side, backing away slowly. There's nothing more they can do, and a levee section right under their feet might collapse at any time. The DWR woman, who is talking to her operations center on a CB radio between commentary for her television audience, reports that a rescue fleet of police and pleasure boats is on its way to Mandeville now to pick up the inmates. If another breach forms on the crews' back side, they'll be stranded on a small section of levee being eaten at either end. I have had my share of experience capsizing in rough whitewater rapids, and I find myself wondering what it would be like to be swept away by the thundering water roaring into that levee breach. Survivable, perhaps, if you're wearing a life vest and you're plucked out in a few minutes; this water was icy snowmelt just a couple of days ago, or was released from the cold bottom of a deep reservoir. But these inmate crews aren't issued life vests, and a lot of them—petty criminals who grew up in cities and ghettoes—probably don't know how to swim.

The reporter and the DWR spokeswoman are as speechless as we are, gathered around the TV; as the camera pans below them, they say nothing. The DWR woman knows what I know: even though 90,000 cubic second-feet of freshwater runoff is pouring into the Delta—the rainy-season runoff from most of the Sierra Nevada—the flow sucked into all these huge sunken islands filling serially amounts to *hundreds of thousands* of cubic feet per second. The

Sacramento River, the San Joaquin, the Mokelumne River, all their connected, meandering Delta channels and sloughs—they're being slurped *backwards*, literally. So is Suisun Bay, which leads to San Pablo Bay, which leads to San Francisco Bay, which meets the ocean at the Golden Gate. A vacuum has been created here, a vacuum that water, obeying gravity, must fill. The water is full of salt.

The reporter in the copter has already figured out the implausible implications of what's happening below. Does all this mean—he asks the DWR spokeswoman—that the great aqueducts feeding Los Angeles and San Diego with half their water supply from the Delta—does this mean, because of the saltwater invasion, that we have to shut them down? he asks. Does it mean that the San Joaquin Valley, the world's most productive reach of irrigated farmland, can no longer irrigate?

The DWR woman hesitates; she's probably afraid to say what it means. The south Delta pumps that fill the Delta–Mendota Canal and the California Aqueduct were shut off hours ago, she says. A DWR boat monitoring salinity with a mobile testing lab found the saltwater profile advancing quickly toward the Clifton Court Forebay, where Delta water enters the pump batteries and, in turn, the aqueducts. But that's no disaster for southern California—not *yet*. San Luis Reservoir, a huge offstream basin in the Coast Range about a hundred miles south, can store a couple of million acre-feet of water and, this being the end of the rainy season, it's nearly full. Several good-sized receiving reservoirs around Los Angeles are filled with northern California water, too. San Joaquin farmers have groundwater below them—many of them do—and a fair portion of the valley gets all or most of its irrigation water from rivers south of the Delta—the Tuolumne, the Merced, the Stanislaus, the Kaweah and Tule and Kern.

Which means what? the reporter wants to know. Whoa! The DWR director has called in somehow—on a protected circuit, I imagine. He's heard his poor flack struggling with answers to questions so explosive they're almost fissionable, and he's decided to take charge himself. This is much too big an issue for a hydrologist to handle.

Now the state's top water official is on the line, and he is reassuring us that Los Angeles and San Diego and their galaxy of suburbs, with a population nearly equal to that of New York State, have enough water stored locally to last for several months. And northern California water pumped through the Delta isn't their only source, not by any means. The Metropolitan Water District of Southern California has 550,000 annual acre-feet from the Colorado River, plus some extra Colorado River water it's purchased from the Imperial Irrigation District. And the City of Los Angeles has its Owens River Aqueduct. Besides those sources, the local rivers—the Los Angeles, the San Gabriel, the Santa Ana—contribute something. Even so, admits the chief, water from northern California and the Delta is half of urban southern California's supply. It also irrigates several million acres of land in the San Joaquin Valley. And, meanwhile, two of East Bay MUD's three aqueducts from the Mokelumne River have severed at a river crossing. All three bores cross the south Delta on pier support—they're not buried, like San Francisco's aqueduct. That's a crisis, but not really a disaster, because the other (and largest) bore seems to be functional. There may be some deformity here and there.

What if we're heading into a severe drought, like 1976 and 1977 or, worse, 1987–92? Aren't we facing unprecedented rationing then? *Unimaginable* rationing, at least in southern California? When can Delta water be pumped into the aqueducts again?

The DWR chief says he has no idea, though he is inclined to agree about the rationing. If we start rationing right now, we can make the water in storage last that much longer. This is an event so far outside our experience, he says, that there's no way to answer these questions. It depends on how many islands end up inundated, and how salty the Delta becomes, and how, long it takes to flush it all back out.

The problem is, you *have* to flush this huge new brackish inland sea all the way out to the ocean, or at least push it all back into Suisun Bay, before you can even think about turning the pumps back on and filling the aqueducts again. And, says the chief, we have some idea what's involved there. In the summer of 1972, Brannan–Andrus, a joined island, flooded when a levee breached, for mysterious reasons—there was no earthquake or flood. That single inundation created a reverse flow powerful enough to suck ocean water far into the Delta. The breach was repaired, and the water was pumped out of the island into the surrounding river channels, but 300,000 acre-feet of water had to be released from upriver reservoirs to flush out the invasive saltwater. The DWR monitors salinity levels at several points in the Delta, from east to west, and tolerance levels are set at each point to protect the flows headed toward the South Delta pumps and aqueducts. If natural outflow from rain and snowmelt aren't enough to keep the salts at bay, then water stored in reservoirs has to be released to push it back to the western edge of the Delta, where it can't contaminate the flows headed into the aqueducts. That becomes a problem in drought years, when the salinity profile advances inland without countervailing natural runoff, and you have to release a great quantity of water from reservoirs—the same water that you want to *conserve*, because, after all, you're in the middle of a drought.

But this situation now has no precedent, nothing even close. Exactly, agrees the reporter. This must be ten times worse than 1972. And how do you even fix these levees when this mass inundation has saturated them and strong shaking has probably weakened them. You could fix them, at enormous cost, could you not, and they could fail again and again, couldn't they? Might it be *hopeless*?

This reporter has caught on fast. The DWR chief rejects that word, *hopeless*. It's a crisis, yes. We're at the end of the wet season. When the snowmelt is over there's no serious contribution from nature, not for months—not until the winter rains return. We hold back a lot of the snowmelt, anyway; it's released by dams only if their reservoirs are nearly full and the snowpack is abnormally deep and the melt gets out of hand. If we use every drop of runoff to flush out the Delta—based on what it took to flush it in 1972—we'll have emptied the reservoirs, probably. There's nothing left for irrigation, for urban water supply.

The DWR chief's last words left everyone stunned. They also gave him a quick exit opportunity. Meanwhile, the helicopter has stationed itself over an obvious slump in a west-facing levee section of Venice Island; the underlying soil has liquefied and a piece of levee—forty feet across? fifty feet?—has collapsed several feet. Water flowing down Potato Slough is barely below the top of that levee section; if it slumps some more, Potato Slough—a channel fed by the Sacramento River—will begin to pour over the top of the embankment, eroding a small breach that could soon become another gaping, thundering, hundred-yard-wide hole. The DWR spokeswoman has found her voice again. She tells us that the Hilton family owns a big piece of Venice Island, and the levees there are well maintained. That's not always the case, she has to admit. Someone is standing down there watching—the farm manager, perhaps?—but there are no work crews in sight. They must be overwhelmed.

The station cuts back to its pair of news anchors in San Francisco, who have been quietly listening to all this while watching the footage on their monitor. As they reappear on the screen, they are looking at each other and shaking their heads. "Unbelievable story," one says. "Here are the latest casualty and damage estimates released by the state . . ."

I go back home, where my family is just waking up. My wife is about to try calling her relatives, which will take hours, and my kids want to talk about the earthquake with their friends. I can't sit. I decide to hook up the boat trailer and head back to Loch Lomond. Even with a small boat, there's got to be some way I can help. My local gas station owner has hooked a generator to his pumps, and he's able to bring up some fuel. Ten gallons maximum. I wait only ten minutes in line—hardly anyone seems to know there's even gas to be had.

We were smart to keep a few hundred dollars in cash in a strongbox. All the banks are closed, and the ATMs are down.

I spend a few minutes watching the rescue effort at the two pancaked apartment buildings in the Canal District; most of the bodies are still inside. With an uneasy stomach, I offer to help dig out, but the offer is spurned; it's too dangerous, I'm told. Local firefighters and inmate crews are working the precarious ruins. At Loch Lomond, the dockmaster laughs at my eighteen-foot contribution to the emergency fleet. He points to a big cruiser that's for sale, cheap. I launch out anyway. Back on the bay, I see the same Coast Guard boats, ferries, navy boats, barges, big cruisers, all running beelines every which way. The sky is full of helicopters and small aircraft. The huge Richmond fire is being quelled; it's billowing smoke, but not as much. Many other fires are still burning, and a ghastly smoke pallor lies over the region. As I near Angel Island, the macabre remnant of the eastern Bay Bridge begins to materialize in its shroud. You can't even see the thickening mackerel-back clouds that the weather reporter just mentioned on the radio; it's going to start raining in a few hours. The Oakland skyline is invisible in the smoke haze; a ghostly delineation of San Francisco is all I can make out. On top of everything, we're half-blind.

What am I doing out here, I ask myself. I want to help, but have no idea how.

More than three thousand people have died since 2:28 yesterday afternoon, but six million others are still alive in the Bay Area. Thousands of structures are wrecked, tens of thousands are damaged—but most aren't. That's one irony of an earthquake in a vast metropolitan area. If a quake levels a little town, the residents—those who survived—might decide to move; they'd abandon the place for somewhere safer, as some Mississippi River towns relocated after the great 1993 floods. But in a built-up area, a place where big earthquakes are certain to cause tremendous damage and loss of life, so much remains intact that rebuilding is a foregone conclusion.

Half of San Francisco burned to the ground in 1906.

What did we do? We rebuilt and told ourselves it wouldn't happen again for a long time. Nearly all of Yokohama and a third of Tokyo—a much larger city than San Francisco in 1906—burned to the ground after the earthquake in 1923. What did Japan do? It put Tokyo right back where it was and made it the most populous city in the world. Another Great Kanto earthquake, according to one estimate I have seen, could become the world's first trillion-dollar natural disaster. Will the Japanese rebuild Tokyo exactly where it is, after that happens? Of course—I just can't imagine how, and with whose money. Just as we will rebuild here. Will Los Angeles relocate, now that it's lost half its water? Where?

Fate, whimsy, ignorance, sheer will—any number of things conspire to put a city where it is. Once it's there, and it's big enough, it's there forever. But if it's in an earthquake zone, as populous California is . . . as populous Japan is, even more so . . . you've created a civilization that becomes drastically expensive to maintain. The reconstruction of the Bay Area will create lots of jobs; it will suck in a tremendous capital investment. It will create—how perverse can things get?—a huge reconstruction boom amid a staggering productivity loss.

But billions and billions of dollars will also come from American taxpayers living elsewhere, and I can already sense what the rest of the country is thinking: *Where does it end?* The peninsula segment of the San Andreas fault could produce an earthquake with consequences similar to this. Is it going to wait another century, or is it going to go in . . . 2008? The last Hayward fault earthquake was in 1868. The San Andreas peninsula segment slipped in 1865. Both quakes were nearly as powerful as what we just experienced, and they were just three years apart.

Then you have southern California and its upcoming earthquakes—it has more faults than we do in the Bay Area. The damage inflicted by an earthquake on the Newport–Inglewood fault could be more expensive than what we've got here. Then add the other disasters: the floods, droughts, fires, and slides.

Give California nationhood. Saw it off from the rest of the United States. That's what they must be thinking in Oneonta, New York, where blizzards are the worst things to come around.

ACKNOWLEDGMENTS

Near the time of his death, Marc Reisner clearly expressed his wish that the manuscript for this book reach print. That would not have been possible without the generous support of many people—his family, friends, and colleagues. Dan Frank, his editor, displayed his customary talents of graceful editing and aesthetic judgment. Joe Spieler, his agent, continued his fierce advocacy for Marc in all affairs literary and offered heartfelt and wise counsel on any matter whenever asked. Both waited patiently for this manuscript.

Gratitude is due to the many people interviewed for this book. No doubt more people than those named in the text generously offered their time to Marc. Thanks to all who lent their professional expertise. In particular, John Eidinger and Bill Lettis provided technical review of the manuscript and graciously assisted me in the final stages of the process. Rich Eisner also spent long hours being interviewed and offered very useful comments at the end of the project. Any errors are the fault of the author and his successor.

Finally, thanks are due to Ruthie and Margot, who first endured their father's preoccupation with the book and then briefly their mother's.

Lawrie Mott

FOR MORE FROM HARD-HITTING JOURNALIST AND ENVIRONMENTALIST MARC REISNER, LOOK FOR THE

Cadillac Desert
The American West and Its Disappearing Water

"A savagely witty history of America's reckless depletion of its water sources." —*Newsday*

"The definitive work on the West's water crisis."
—*Newsweek*

"A highly partisan, wonderfully readable portrayal of the damming, diverting, and dirtying of western rivers."
—*The Washington Post Book World*

"Intelligent, provocative, and compulsively readable."
—*Chicago Sun Times*

The story of the American West is the story of a relentless quest for a precious resource: water. It is a tale of rivers diverted and dammed, of political corruption and intrigue, of billion-dollar battles over water rights, of ecologic and economic disaster. In *Cadillac Desert* Marc Reisner writes of the earliest settlers, lured by the promise of paradise and of the ruthless tactics employed by Los Angeles politicians and business interests to ensure the city's growth. He also documents the bitter rivalry between two government giants, the Bureau of Reclamation and the U.S. Army Corps of Engineers, in the competition to transform the West. Based on more than a decade of research, *Cadillac Desert* is a stunning exposé and a dramatic, intriguing history of the creation of an Eden that may only be a mirage.
ISBN 0-14-017824-4